Production and Commercialization of Insects as Food and Feed

Francesco Montanari • Ana Pinto de Moura
Luís Miguel Cunha

Production
and Commercialization
of Insects as Food and Feed

Identification of the Main Constraints
in the European Union

 Springer

Francesco Montanari
School of Law
Universidade Nova de Lisboa
Lisbon, Portugal

Ana Pinto de Moura
Department of Sciences and Technology
Universidade Aberta
Lisbon, Portugal

Luís Miguel Cunha
Faculty of Sciences
University of Porto
Porto, Portugal

ISBN 978-3-030-68408-2 ISBN 978-3-030-68406-8 (eBook)
https://doi.org/10.1007/978-3-030-68406-8

This Springer imprint is published by the registered company Springer Nature Switzerland AG
The registered company address is: Gewerbestrasse 11, 6330 Cham, Switzerland

Preface

This book *Production and Commercialization of Insects as Food and Feed: Identification of the Main Constraints in the European Union* is the result of research work developed in the context of the MSc in Food Consumption Sciences of *Universidade Aberta* (Open University of Portugal) by Dr Francesco Montanari and supervised by Ana Pinto de Moura (Assistant Professor at *Universidade Aberta*) and Luís Miguel Cunha (Associate Professor at the University of Porto, Faculty of Sciences).

Both Prof. Luís Miguel Cunha and Dr Ana Pinto de Moura have been working in the evaluation of consumers' acceptance of edible insects since 2013. When Dr Francesco Montanari started to design his MSc dissertation, back in 2018, they challenged him to identify the main legal barriers in the European Union (EU) to the adoption of insects as food and feed. This would allow to combine their research interests with the vast experience of Dr Francesco Montanari in the area of EU food law.

Forecasts point out an exponential growth in the global population, which raises concerns and doubts over the ability of the current agri-food production systems to meet food demand in the medium and long term. Such a prospect has led international organisations and the scientific community to raise awareness about, and call for, the need to identify additional sources of food to feed the world. From this perspective, insects qualify as a suitable and more environmentally friendly alternative to meat and other foods that are source of animal proteins. Notwithstanding that, the uptake of the production and commercialisation of insects as food has been facing regulatory hurdles, consumer scepticism and rejection in many markets. This is particularly true in the context of Western societies in which insects do not always constitute part of the local traditional diets.

Against this background, the proposed work intends to analyse and discuss the regulatory state of the art for the production and commercialisation of insects as food and feed in the EU. Over the last few years, the EU has been taking concrete legislative steps with a view to opening up its market to insect-based food products. Yet, some key regulatory constraints still exist today, which ultimately prevent the industry sector from growing, consolidating and thriving, and its players from competing on a level playing field.

Currently, the main regulatory constraints in the EU for insects as food include the fragmentation of the EU market as a result of the adoption of different policy solutions by EU member states as regards their status as novel foods and the length and complexity of the relevant authorisation procedures. Also, ad hoc safety and quality requirements tailored to the needs and specificities of the insect food sector are currently missing at EU level. Conversely, the regulation of insects as feed in the EU is at a much more advanced stage, although further regulatory steps would be desirable to ensure consolidation and expansion of this market segment.

In spite of the regulatory constraints and loopholes identified, drawing a comparison between the EU and other countries for which insects likewise constitute a novelty food and feed wise, the EU market presents a higher degree of openness as well as greater potential and maturity from a regulatory viewpoint. In the long run, this can make the EU a global leader for insects as food and feed both as a market and as a standard-setting body.

Overall, the present work constitutes the first attempt at providing a comprehensive overview of the evolution over time and of the current state of the art of the regulatory framework for insects as food and feed in the EU, based on a multidisciplinary approach that combines science, policy and law and that reflects the academic background of the authors. From this baseline, the work then contains the following innovative and distinctive elements:

- It proposes a legislative roadmap which the EU should follow in order to make its regulatory framework fit for insects as food and feed in the long term.
- It provides a detailed comparison between the current EU legal framework and other regulatory systems of Western countries – including a few countries which have not been studied to date (e.g. Switzerland and Brazil) – with a view to singling out the market(s) which are better equipped to address the production and commercialisation of insects as food and feed.
- It provides an updated overview of the overall market and of European consumers' perspectives on the use of insects as food and feed.

We believe that this book will appeal to a large audience – from companies to young entrepreneurs acting in the EU market of alternative animal feed and edible insects, as well as researchers and students of all ages – addressing novel foods, alternative protein sources and food supply sustainability through the use of insects as food and feed.

We would like to dedicate this book to our families, particularly to Duarte, José and António.

Lisbon, Portugal   Francesco Montanari
                                                              Ana Pinto de Moura
Porto, Portugal   Luís Miguel Cunha
26 November 2020

Contents

About the Authors

Dr. Francesco Montanari is a lawyer specialised in the agri-food chain and an Integrated Researcher at CEDIS – Research and Development Centre on Law and Society of the School of Law of the *Universidade Nova of Lisbon* (https://cedis. fd.unl.pt/en/). He holds a PhD in EU law from the University of Bologna and a MSc in Food Consumption Sciences from the *Universidade Aberta*. His main areas of interest and research revolve around food innovation (including novel foods, insects, dairy and meat substitutes), food labelling, nutrition and healthy diets, official controls and international trade in agri-food products. Since 2013, he lectures food law in training courses organised by the European Commission for the staff of national competent authorities. E-mail: francesco.montanari@arcadia-international.net

Dr. Ana Pinto de Moura is an Assistant Professor at the *Universidade Aberta* (https://www2.uab.pt/departamentos/DCT/detaildocente.php?doc=30) and the Director of the online MSc course on Food Consumption Sciences, since 2005. She is also an Integrated Researcher at the GreenUPorto – Research Centre for Sustainable Agrifood Production (https://www.fc.up.pt/GreenUPorto/en/), where she integrates the research group working on consumer perception and sensory evaluation. Her main areas of research deal with the use of qualitative research methods to unravel determinants of food choice with a focus on healthy and sustainable eating. E-mail: apmoura@uab.pt

Prof. Luís Miguel Cunha is an Associate Professor at the Faculty of Sciences, University of Porto (https://sigarra.up.pt/fcup/en/FUNC_GERAL.FORMVIEW?p_codigo=405494) and the Director of the MSc course on Consumer Sciences and Nutrition, since 2007. He is also an Integrated Researcher at the GreenUPorto – Research Centre for Sustainable Agrifood Production (https://www.fc.up.pt/GreenUPorto/en/), where he coordinates the research group working on consumer perception and sensory evaluation. His main areas of research deal with the application of fast sensory profiling techniques for food products, dynamic profiling, food-evoked emotions and food choice criteria. Within this framework, since 2013, he has focussed part of his research efforts on the evaluation of the use of insects as food and feed. He is a member of the European H2020 innovation action SUSINCHAIN – Sustainable Insect Chain (2019–2023). E-mail: lmcunha@fc.up.pt

Acronyms

AFSCA	Agence fédérale pour la sécurité de la chaîne alimentaire (Belgium)
ANSES	Agence nationale de sécurité sanitaire de l'alimentation, de l'environnement et du travail (France)
ANVISA	Agência Nacional de Vigilancia Sanitária (Brazil)
ASBRACI	Associação Brasileira dos Criadores de Insetos
B2B	Business-to-business
B2C	Business-to-consumer
BIIF	Belgian Insect Industry Federation
BSE	Bovine spongiform encephalopathy
CFIA	Canadian Food Inspection Agency
CFR	Code of Federal Regulations (USA)
DGAV	Direção Geral de Alimentação e Veterinária (Portugal)
EC	European Commission
ECJ	European Court of Justice
EFSA	European Food Safety Authority
EP	European Parliament
EU	European Union
FAO	Food Agricultural Organization
FDA	Food and Drug Administration (USA)
FSANZ	Food Standards Australia and New Zealand
GHGs	Greenhouse gases
GRAS	Generally recognized as safe
HACCP	Hazard Analysis and Critical Control Points
IPAA	Insect Protein Association of Australia
IPBES	Intergovernmental Science-Policy Platform on Biodiversity and Ecosystem Services
IPIFF	International Platform of Insects for Food and Feed
MAPA	Ministério da Agricultura, Pecuária e Abastecimento (Brazil)
MIPAAF	Ministero delle Politiche Agricole, Alimentari e Forestali (Italy)
NACIA	North American Coalition for Insect Agriculture (US)
NGO(s)	Non-governmental organisation(s)

OIE   World Organisation for Animal Health
OSAV   Office fédéral de la sécurité alimentaire et des affaires vétérinaires
            (Switzerland)
PAPs   Processed animal proteins
RASFF   Rapid Alert System for Feed and Food
SPS   Sanitary and Phytosanitary Agreement
US/USA   United States/United States of America
WTC   Willingness to consume
WTO   World Trade Organization

List of Figures

List of Tables

Chapter 1
Introduction

The use of insects as food and feed has gained unprecedented prominence at international level both in scientific and policy settings over the last decade, mainly owing to the role that such animals may play in feeding the world and reducing the environmental footprint of the modern agri-food production systems.

Indeed, in accordance with the Food Agriculture Organization (FAO), the world's population is expected to reach approximately 10 billion people by 2050, representing a growth of 38% compared to 2012. This scenario puts inevitably lot of pressure on the current models of agricultural and food production; at the same time, it raises doubts over their capability to meet the higher level of food demand, notably of animal-based proteins, which should result from the increase in the world's population in a way that is environmentally sustainable.

It is against this background that international organisations, the scientific community as well as non-governmental organisations (NGOs) have embarked, already for some years now, in the quest for alternative protein sources for human and animal consumption. In this context insects – perhaps together with algae, seaweed and duckweed – are amongst the protein sources that have sparked more interest from scientists, industry players and policy-makers alike on a global scale, in light of their multi-fold applications in food and feed production and inherent business potential (Lamsal et al. 2019).

It is a fact though that the production and commercialisation of insects as food and feed have not developed uniformly across the planet as of yet. While insects form part of the traditional diets of the population of several countries in Asia, Africa and Latin America and are therefore produced, marketed and generally accepted by local consumers in such regions, the same cannot be said with reference to North America and Europe, for instance. Likewise, the use of insects as feed can be subject to different requirements across countries or regions, having regard to the specific food safety, environmental and/or animal health objectives pursued by local feed legislation.

© The Author(s), under exclusive license to Springer Nature
Switzerland AG 2021
F. Montanari et al., *Production and Commercialization of Insects as Food and Feed*, https://doi.org/10.1007/978-3-030-68406-8_1

This being said, the present work aims at providing a comprehensive and critical review of the regulatory framework applicable to insects as food and feed in the European Union (EU). As one of the largest global markets, the EU has been, in fact, gradually opening up to the production and commercialisation of insects as food and feed over the last few years, by eliminating some key regulatory barriers in this area. Nevertheless, some regulatory hurdles still remain in place at EU level, especially as far as insects as food are concerned. For this reason, while the scope of the present work covers, in principle, insects intended for both food and feed, it has been deemed appropriate, wherever relevant, to give more prominence to aspects and issues linked with human consumption.

Under that premise, in order to fully apprehend the scientific and policy context in which the EU has been taking steps to legitimise the use of insects in human and animal nutrition, Chap. 2 of this work illustrates the main reasons that have led scientists and policy-makers to advocate for the integration of insects in the food and feed chain, giving account of their technical implications and current limitations.

Chapter 3 has instead the objective to provide a characterisation of the current insect market for human and animal nutrition at global and European level, in terms of size, main industry players and potential.

Following that, Chap. 4 portrays the current regulatory context of few other selected Western countries – namely the United States of America (USA), Canada, Australia, Switzerland and Brazil -in which the use of insects as food and feed has been subject to recent policy discussion and/or legislative measures similarly to the EU, with the objective to subsequently draw a regulatory comparison between the EU and such countries.

The analysis of the regulatory evolution undergone by the EU so far and of the barriers and obstacles that still remain for the production and commercialisation of insects as food and feed is finally dealt with in Chap. 5.

Chapter 2
Insects as Food and Feed

2.1 Insects and the Agri-Food Chain

Insects currently represent 70–75% of all animal species living on earth (Katayama et al. 2007). The class of invertebrates *Insecta* of the phylum *Arthropoda* consists of over one million known species, while approximately five million species are in fact thought to exist in total. There are currently about 23 different insect orders, amongst which *Coleoptera* (beetles), *Diptera* (flies), *Hymenoptera* (bees, ants and wasps), *Lepidoptera* (butterflies and moths) and *Orthoptera* (crickets and grasshoppers) (Evans et al. 2015).

In recent years, public interest and scientific research focussing on the potential use of insects as a source of food and feed have been steadily growing worldwide, largely because of the social and economic concerns associated with the demographic increase that analysts expect to take place over the next few decades as well as taking into account the environmental impact of the current agri-food production systems (Yen 2009; van Huis 2013, 2020a) (see further Sects. 2.3 and 2.4).

If today it is to some extent reasonable to expect that most international and national legal systems alike will progressively come to integrate insects under the existing set of rules governing the production and the commercialisation of food and feed, it cannot be neglected though that insects play other roles and functions in the current organisation and management of the agri-food chain.

On the one hand, some insect species represent a threat to agricultural crops, livestock and their products and/or to biosecurity. For instance, the Asian black hornet (*Vespa velutina*)- an insect which is native of Southern-Eastern Asia though now well-established in Europe, including in the Iberian Peninsula- may cause considerable losses to apiculture as it is a predator of honeybee foragers (López et al. 2011). Also, several species of mosquitos can transmit the virus of the genus *Phlebovirus,* order *Bunyavirales,* family *Phenuiviridae,* which may cause the Rift Valley Fever. This is a sickness that was detected for the first time in the homonymous

F. Montanari et al., *Production and Commercialization of Insects as Food and Feed*, https://doi.org/10.1007/978-3-030-68406-8_2

valley in Kenya in 1930 on a sheep farm and, then, in few other countries in Africa, the Middle- East and Europe, and that can be fatal for food-producing animals (e.g. goats, sheep, cattle) and humans (World Organisation for Animal Health (OIE) 2020). From this standpoint, insects are therefore regarded and increasingly regulated, nationally and internationally, as pests, invasive species or vectors of diseases (van Huis et al. 2013; van Huis 2020a).

On the other hand, some insect species are essential for the equilibrium and dynamics of certain ecosystems. Insofar as they make so that nature takes its own course or contribute to speeding up its processes, they provide regulating and supporting services, in accordance with the classification elaborated by Noriega et al. (2018).

This is notably the case of bees and other pollinators such as flies, ants, and certain butterflies, amongst others, which, by fertilising plants, make them produce flowers, fruits and seeds (Rader et al. 2016). Besides that, certain insects pertaining to predatory or parasitoid species may be useful tools in agriculture, notably to ensure biological control of plants' pests. Indeed, the use of insects as biological control agents is a practice that goes back in time, with the first documented case, where ants were used to control pests in citrus orchards, occurring in 300 AD. A more recent success story consisted in the use of an imported ladybird (*Rodolia cardinalis*) and a dipteran parasitoid (*Cryptochaetum iceryae*) for the control of the cottony cushion scale (*Icerya purchasi*), another citrus pest, at the end of the XIX century, first in California, and, subsequently, in other regions and countries (Bale et al. 2008). Currently, in economic terms, the role of insects as pollinators has been estimated by the Intergovernmental Science-Policy Platform on Biodiversity and Ecosystem Services (IPBES) in around 235–577 billion US$ annually (IPBES 2016), whereas natural biological control performed by certain insect species would be worth 400 billion US$ per year (van Lenteren 2006).

Finally, the role that certain inspect species play namely as decomposers of organic matter (e.g. litter, wood) and, therefore, as contributors to the preservation of specific ecosystems (e.g. freshwaters) has been generally well documented in scientific literature (Kampichler and Bruckner 2009). However, further studies are still being conducted nowadays, in particular, as a result of the adoption of international and national policies that promote a model of circular economy. This is in fact particularly relevant in the case of certain insect species intended for human consumption and animal nutrition, which, by being fed on food waste, can ensure its valorisation while avoiding the environmental damage ensuing from traditional solutions for waste disposal (e.g. landfilling) (Varelas 2019) (see also further Sect. 2.4).

2.2   Insects as Food: History, Culture and Traditions

The human practice of eating insects, otherwise known as 'entomophagy', is not a recent phenomenon or the latest passing fad (Bodenheimer 1951; Dobermann et al. 2017). Archaeological evidence indicates that insects were originally part of the human diet: *Australopithecus robustus* Broom 1938, for instance, would have use bone tools to dig into termite mounds over 500,000 years ago (Sutton 1995; van der Merwe et al. 2003). Several references to entomophagy can be found also in the Bible (Evans et al. 2015; Schrader et al. 2016): the manna described in the Old Testament, for instance, would be nothing else than the sweet excretion of the mealy bug (*Trabutina mannipara*) that feeds on tamarisk (Medeiros Costa Neto 2013).

However, even in the more recent history, one can easily find proof that insects have traditionally formed part of the diets of local populations inhabiting tropical regions (e.g. Africa, Asia, Latin America and Oceania) as a primary or complementary source of animal proteins and fats as well as of vitamins and minerals and that, in many instances, they are still part of those diets. In the social and cultural settings under exam, therefore, edible insects would not necessarily constitute emergency foods to resort to, for example, in situation of starvation (Bodenheimer 1951; De Foliart 1992a, b; Medeiros Costa Neto 2013; van Huis et al. 2013; Lesnik 2017; Raheem et al. 2018; van Huis 2020a).

In most developing countries, low-income indigenous or rural population groups are typically the largest consumers of edible insects. However, notable exceptions to this pattern exist where insects are considered as delicatessen or luxury food, accordingly priced and eaten by high- and middle-income population groups (e.g. ant pupae known as '*escamoles*' in Mexico and deep-fried insects sold in street markets in Thailand) (Conconi 1982; Chen et al. 1998). Depending on the local cultural and cooking traditions, insects may be prepared to be consumed in different ways, including raw, sun-dried, boiled, steamed, roasted or deep-fried; moreover, depending on the specific insect species, all instars or only some of them (e.g. eggs, nymphs, pupae, larvae, adult etc.) may be explored for such a purpose (Bodenheimer 1951; De Foliart 1992b; Nonaka 2009; Medeiros Costa Neto 2013; Raheem et al. 2018).

Although edible insects are part of the local diets in many countries, relevant studies indicate that, in some instances, the uptake of Western-type lifestyles, as an effect of the current globalisation, migratory fluxes and urbanisation, has contributed to the introduction of substantial changes to traditional dietary patterns (Illgner and Nel 2000; Gracer 2010; Evans et al. 2015).

By way of an example, Yen (2009) refers that the arrival of the Europeans in Australia led to a sharp decline in the consumption of edible insects amongst the native Aborigines. In fact, in the long term, this resulted in the adoption of European diets by those indigenous populations with the associated health burden that such diets generally involve (e.g. diabetes and cardiovascular diseases). In light of this, there are currently doubts that one day insects as such could turn into ordinary food in the country, whereas their commercialisation as animal feed, in food supplements or in ethnic restaurants would bear greater potential. In the case of Thailand, Chen et al. (1998) note that while, historically, insects were eaten by the poor and predominantly in rural areas, the progressive industrialisation of the country and urbanisation of the population that occurred in the second half of the last century fuelled the perception of insects by urban and educated people as a food of the poor and, as a result, propelled a generalised reluctance vis-à-vis their consumption. Nonetheless, some insect species are still served occasionally as luxury or gourmet foods in the finer restaurants or hotels of the largest cities. Even in China, which is one of the countries with the highest number of edible insect species worldwide, the improvement of living conditions has engendered a certain loss of traditional knowledge in this area across the country, although the custom is still very much rooted in certain provinces (e.g. Yunna) (Chen et al. 2009).

Conversely, entomophagy is at present far less rooted in Western societies and this for a number of different historical and/or cultural reasons (see further Sect. 2.7). Indeed, it can be said that today the practice of eating insects can trigger diverse reactions amongst Western consumers: this includes fear of the unknown and risk avoidance (commonly known as *'food neophobia'*), disgust, abhorrence but also curiosity (De Foliart 1999; Yen 2009; Cunha et al. 2014; Caparros Megido et al. 2016; Tan et al. 2016; La Barbera et al. 2018). In this respect, it is worth noting that the term 'entomophagy' was coined by Westerners in the second half of the nineteenth century to describe the (peculiar) habit of eating insects observed in certain human populations. Yet, equivalent terms do not exist in all languages and cultures and notably where edible insects are traditionally consumed. At the time of its coinage, the word 'entomophagy' was therefore given a pejorative connotation by implicitly referring to a social behaviour deemed inappropriate – a connotation which has essentially remained unaltered over time and which still reflects the Western bias toward this practice (Evans et al. 2015).

Notwithstanding that, research shows that entomophagy has been practiced in the past in all continents, except in Antarctica (Bodenheimer 1951). Accordingly, Schrader et al. 2016 refer that in Northern America (north of Mexico including Greenland) entomophagy was relatively common amongst indigenous people and, in certain cases, even amongst colonists. In this context, the most memorable case is probably that of the Mormon cricket (*Anabrus simplex*). This katydid was responsible for the destruction of the yields of the Mormon settlers in Utah in 1848. The Ute, a native local tribe, used to collect this insect from the wild, dry and crush it until a became a powder to make prairie cakes, which they offered to the settlers to help them go through winter.

Rumpold and Schlüter (2013) point out that limited data are available with regard to entomophagy in the European continent. Yet, the diet of both ancient Greeks and Romans featured insects often as delicacy food, including grasshoppers, cicadas and a bug (*Lucanus cervus*) which was purposefully fattened for human consumption. Whilst the practice of eating insects appears to have started its decline following the fall of the Roman Empire, in modern history there are still quite a few documented accounts of entomophagy, namely in time of starvation (e.g. during the Irish famine in 1688) or during insect outbreaks (e.g. locusts in Germany in the late seventieth century) (Bequaert 1921; Bodenheimer 1951; Payne et al. 2019). More recently, Dreon and Paoletti (2009) refer that some insect species and their products had been traditionally consumed by local populations in the North-East region of the Alps in Italy until 1980s, whilst today only one insect byproduct – that is ingluvies of lepidoptera of the genus *Zygaena* and *Syntomis*- would still be eaten in some villages. Likewise, in certain European countries some traditional food products contain insects. One of the best-known examples is the

Sardinian cheese 'Casu Marzu'. This traditional product contains fly larvae which are commonly eaten alive with the cheese and which contribute to the fermentation process and the organoleptic characteristics of the food. Interestingly, while the commercialisation of this local cheese was banned in 1962 owing to a growing aversion against a product that symbolised poverty and backwardness, it was subsequently included in the national list of protected traditional agri-food products. Similar products can be found in other parts of Italy, France, Croatia and Spain (Manunza 2019; Payne et al. 2019; Ministero delle politiche, alimentari e forestali (MIPAAF) 2020). Besides that, in accordance with recent data collected and analysed by the European Food Safety Authority (EFSA), the EU risk assessor, few insect species have been farmed for human consumption for quite some time in the EU, including the house cricket (*Acheta domesticus*), the yellow mealworm (*Tenebrio molitor*), the lesser mealworm (*Alphitobius diaperinus*) and the migratory locust (*Locusta migratoria*), while others might be occasionally imported from non-EU countries (EFSA 2015; Belluco et al. 2017).

This being said, one also often forgets that insects provide us today with food and ingredients. Bees are probably one of the great sources of human food, providing not only honey and propolis, but also beeswax, which is widely used by the food industry as an additive (E901) in different product categories (e.g. confectionery, snacks, fruits and vegetables etc.) and with different functions (e.g. glazing agent, thickener, carriers for colourings). Likewise, carmine – also known as carminic acid or Natural Red 4 – is a permitted natural food colouring (E120), which is extracted from the insect species *Dactylopius coccus* and nowadays widely used also by the cosmetic industry. Other than that, the potential of insects as food and elements of macro- and micro-agricultural schemes has been explored in relation to specific non-conventional settings, including space and space vehicles (Katayama et al. 2007).

Against this background, it has been recently estimated that there are currently over 2,000 species of edible insects worldwide with entomophagy being practiced in 80% of the countries. In this context, beetles, caterpillars and ants, bees and wasps rank as the three most consumed categories (Jongema 2017).

2.3   Insects and Food and Feed Security

While scientific interest in entomophagy and, more in general, in the potential of insects as food and feed has been present and somehow consistent over the last thirty or forty years, it is only during the last decade that this issue has gained significant public attention (van Huis 2020b). This is largely the result, on the one hand, of effective raising-awareness actions by some entomologists in Europe and in the US (e.g. De Foliart, Meyer-Rochow) (Payne et al. 2019) and, on the other hand, of the research work on edible insects initiated by FAO back in 2003 (Monteiro et al. 2020), which ultimately led to the publication, in 2013, of the book '*Edible insects: future prospects for food and feed security*' (van Huis et al. 2013). This widely cited work constitutes the very first comprehensive assessment of the economic, environmental, health and social benefits which the use of insects for human and animal consumption may bring to the current agri-food systems and to our society as a whole. The relevance of insects as food and feed would be particularly justified in the present century owing to a number of known, emerging or forecast global trends and factors, which include, amongst others, a growing world's population and the increase in food demand which should ensue from that, rising costs for animal proteins, environmental degradation and depletion of natural resources (Yen 2009; van Huis et al. 2013).

As regards the current and the expected growth in the global population, this poses serious concerns in the first place in terms of food security. The latter, in accordance with the definition universally agreed at the FAO World Food Summit in 1996, means ensuring that «*all people, at all times, have physical and economic access to sufficient safe and nutritious food that meets their dietary needs and food preferences for an active and healthy life*». Indeed, FAO estimates that the world's population will pass from 7.1 billion people in 2012 to approximately 9.8 billion in 2050, thus pressuring agri-food system to produce 50% more by 2050 to keep up with food demand, notably of meat and, more in general, of animal proteins (FAO 2017).

While, according to FAO, an increase in agricultural output can be achieved in the medium term, this will inevitably require considerable efforts and resources, including more agricultural land, more livestock and, thus, more animal feed (e.g. cereals and other protein-rich plant commodities) and other agricultural inputs (e.g. water, fertilisers, pesticides), resulting, in all likelihood, in an undesired impact on the environment, availability of natural resources and biodiversity (van Huis 2013; Kirova et al. 2019) (see further Sect. 2.4).

From this perspective, together with other food (e.g. algae, seaweed, rapeseed) and food innovations under development (e.g. meat grown *in vitro*), edible insects have been identified, and have gradually emerged, as suitable alternative protein sources to conventional food for the future (Aarts 2020; da Silva Liberato and Melo de Vasconcelos 2020; Cardoso et al. 2020). High feed conversion rates – that is the capacity to transform feed into edible meat – and the reduced environmental impact associated with insect farming compared to traditional food-producing animals (see further Sect. 2.4) are generally the arguments which are most commonly put forward to justify insects' potential and call for a greater integration of these animals in the current agri-food systems and human diets (van Huis 2015). From a food security point of view, it is worth noting that edible parts of insects amount to 80–100% depending on the species, while for other livestock this percentage is significantly lower on average (notably, 55% in case of bovines and 40% for poultry and swine) (van Huis 2013; Dobermann et al. 2017).

Likewise, a sharp increase in the total world's population raises concerns in relation to feed security. As referred above, more feed will be necessary to rear food-producing animals that can provide humans with enough food and proteins, thereby putting increased pressure on agriculture and the environment. From this standpoint, acceptability of insects as animal feed is high as, on the one hand, some species form already part of the diet of certain food-producing animals (e.g. fish, poultry, and pigs), while, on the other, their use would contribute towards the application of increasingly natural farming practices (Rumpold and Schlüter 2013; Lähteenmäki-Uutela et al. 2018).

In addition to that, stepping up the use of insects for animal nutrition may entail also some environmental and economic advantages. In particular, insects as feed could serve to partially replace certain feed preparations currently used by the feed industry, but which are becoming increasingly scarce and, therefore, more expensive (Lähteenmäki-Uutela et al. 2018; Lamsal et al. 2019; Gasco et al. 2019; Cardoso et al. 2020). This is the case of fishmeal and oil whose demand has increased exponentially worldwide over the last decades, in particular due to the uptake of the aquaculture industry, and whose production poses environmental concerns for the overexploitation of wild fishery stocks involved in it. It is also the case of soybean meal and oil, which, besides being produced at the expenses of wild forests and other natural environments, are considered to possess lower nutrition quality (e.g. imbalances between essential and non-essential amino acids, anti-nutritional factors, low palatability etc.) as opposed to insect meals (van Huis 2013, 2015).

Furthermore, although research is still at an early stage in this area, some insect species administered as or with feed have shown not to affect negatively animal growth (Rumpold and Schlüter 2013). Some have also proved to be effective agents against common pathogens that are a threat to the health of certain farmed animals (e.g. pigs, broiler chicks and poultry in general). From this perspective, the use of insects in animal nutrition could be a potential and more sustainable option to the administration of antibiotics and, as such, contribute towards their reduction (van Huis 2013, 2020a).

Overall, the black soldier fly (*Hermetia illucens*), the house fly (*Musca domestica*), the silkworm (*Bombyx mori*), the lesser mealworm (*A. diaperinus*) and several grasshopper species are amongst the most promising insect feed replacement alternatives that are currently being studied (van Huis 2013, 2015).

2.4   Insects and the Environment

In addition to being one of the possible solutions to ensure to food security in the long run at global level, the current public and private interest in insects as food and feed is also driven by sustainability considerations as the environmental footprint of such animals would be better than that of conventional livestock.

In fact, over the last few decades the long-term sustainability of the modern agri-food production systems has been increasingly questioned by the scientific community, civil society and public institutions alike. Agriculture, in particular, is regarded as one of the economic sectors which most contributes towards the current environmental burden of the planet. Agriculture heavily relies on key natural resources in order to be able to function (notably, around 70% of water and 40% of land surface which are available globally) and is responsible for 21% of all greenhouses gases (GHGs) emissions. Although the individual contribution and share of responsibility by each continent can be significantly different, globally agriculture significantly contributes to phenomena of environmental degradation such water scarcity, global warming, deforestation, biodiversity loss and topsoil erosion (van Huis et al. 2013; Dobermann et al. 2017; Kirova et al. 2019).

Amongst the various activities of which agriculture consists of, livestock production stands out as one of the most threatening ones for the environment. Besides occupying 70% of all agricultural land, it is responsible for 72–78% of all agricultural GHGs emissions globally. Enteric fermentation of ruminants, manure-related emissions and low feed conversion efficiency of conventional livestock are generally singled out as the main reasons behind the high environmental impact caused by rearing food-producing animals (Oonincx and de Boer 2012; van Huis 2020a).

Against this background, what would be then the main environmental advantages associated with the rearing of insects for human consumption and animal nutrition? To date few scientific studies have been conducted in this area, which, overall, seem to indicate that farmed insects might represent a more environmentally friendly option (van Huis and Oonincx 2017).

In the first place and in comparative terms, insects would require a more limited amount of feed than conventional livestock to produce meat. According to van Huis (2010), in order to produce 1 kg of meat, 25 kg, 9.1 kg and 4.5 kg of feed are required for beef, pork and chicken, respectively, while only 2.1 kg are needed in the case of crickets. The more efficient feed conversion rates shown by insects are related to the fact that these animals are cold-blooded and rely on their environment to control metabolic processes (van Huis et al. 2013). Other than that, insect rearing would generally need less land and water for operating large- scale (mini-livestock) productions: this is particularly apparent when certain insect species (e.g. mealworms, crickets) are compared to cattle as it has been estimated that the former would need 5 and 3 times less land and water, respectively, than the latter (Dobermann et al. 2017).

Likewise, preliminary studies suggest that the overall carbon footprint of certain edible insects (e.g. *T. molitor*, *A. domesticus* and *L. migratoria*) would score better than that of large cattle and be similar to that measured for poultry. In this respect, consideration should also be given to that the improvement of the carbon footprint of insects appears to be highly dependent on the feed which is administered to them. From this perspective and that of a circular economy, feeding insects, for instance, on non-utilised organic waste would seem a better option than reliance on traditional feed (van Huis 2013, 2020a, b; Dobermann et al. 2017).

In contrast, the number of insects required to prepare a meal and their high mortality rates compared to other food-producing animals when bred in captivity are amongst the arguments put forward by some authors against the idea that farming insects is a sustainability-driven business model beyond any criticism (van Huis 2020c).

All in all, as insect farming expands globally, more studies based on insect lifecycle assessment and other relevant environmental parameters are needed to better understand how the environmental benefits described above can effectively be maintained while production is gradually scaled up and increasingly regulated (Dobermann et al. 2017; van Huis 2020a).

2.5   Insects: Food Safety and Quality

As any other food, the production and the commercialisation of edible insects can involve hazards and, therefore, potential risks that may be relevant from a public health perspective. Of course, insect species whose ingestion is known to be potentially toxic for humans, due to the presence of certain compounds, as a result of both synthesis and bio-accumulation (i.e. cryptotoxics such as beetles) or only of synthesis (i.e. phanerotoxics such as bees and ants), are not classified as edible (Belluco et al. 2013; Dobermann et al. 2017).

Furthermore, several scientific studies conducted in recent years have contributed towards a better understanding of food safety implications of rearing, processing and commercialising edible insects and, in doing so, to the identification of possible risk mitigation and management measures to be adopted by insect business operators.

As far as microbiological risks are concerned, overall, research has shown that certain edible insects, including flies and beetles, can be infected with pathogens such as *Salmonella* and *Campylobacter*, following close contacts with infected flocks. In turn, such insects are also a possible vector for the transmission of the infection to livestock. As the insect gut microbiota has been identified as the main source of the microbiological contamination, some studies have suggested that starving insects prior to slaughter may reduce pathogens in the gut and the risk associated with them (Belluco et al. 2013; Dobermann et al. 2017).

Other than that, with regard to processing, some studies indicate that edible insects should not be treated differently from other foods of animal origin, requiring likewise the application of good hygiene practices (e.g. washing and thorough heating) (Grabowski and Klein 2016). Along the same lines, Poma et al. 2017 concluded that insect consumption would not involve additional microbiological and contaminant hazards than consuming other types of meat, as long as comparable levels of good manufacturing practices are guaranteed. Nonetheless, development of microbiological hazards during storage of edible insects needs further investigation to identify the appropriate shelf-life for such products and the relevant conservation conditions to be observed to ensure safety throughout that period of time (Rumpold and Schlüter 2013; Dobermann et al. 2017).

Amongst food contaminants, heavy metals are a source of concern for edible insects. Several studies have shown that bioaccumulation of certain heavy metals (e.g. cadmium and lead) through diet in insect bodies, besides impairing insect development, can exceed safety limits and therefore constitute a food safety risk (Belluco et al. 2013; van Huis 2020a). With regard to pesticides, according to Rumpold and Schlüter 2013, edible insects collected from the wild generally contain residues of plant protection products whenever they have fed in treated areas.

Nonetheless, more recent experimental studies on the relation between pesticides and edible insects, in which the animals were fed with feed spiked with such substances, did not point out to any adverse impact on insect development, on the one hand, and bioaccumulation in the insect body, on the other hand. Similar findings have also emerged with regard to contamination occasioned by mycotoxins, notably by aflatoxins (van Huis 2020a), in spite of prior evidence indicating that edible insects could cause food poisoning for the presence of such contaminants (Schabel 2010).

Whereas allergenic reactions caused by stings of insects or occupational handling of insects as feed and food (i.e. inhalant allergies) have been generally well documented in scientific literature, over the last five or six years a great deal of research has been devoted to the allergenic properties of insects intended for human consumption (Ribeiro et al. 2018). Beforehand studies on this topic were rather limited in number with differences in food traditions being considered as a factor that might condition the prevalence of food allergies (Belluco et al. 2013).

The latest scientific research in this area has confirmed that several edible insects – including the house and field crickets (*A. domesticus* and *Gryllus campestris*), the yellow mealworm (*T. molitor*), the desert locust (*Schistocerca gregaria*) and silkworm pupae (*B. mori*) – contain proteins (notably, tropomyosin and arginine kinase) that can induce allergic reactions in individuals who are sensitive to crustaceans, a phenomenon otherwise known as 'cross-reactivity' or 'co-sensitisation' (Ribeiro et al. 2018; van Huis 2020a). Conversely, cross-reactivity in subjects who are allergic to house dust mites is still being studied and debated (Ribeiro et al. 2019).

In spite of the difficulties to establish the actual prevalence of allergic reactions due to the current uneven uptake of entomophagy across the world and to many cases that may go unreported (Ribeiro et al. 2018), these specific food safety implications ultimately suggest that proper food information needs to be provided to final consumers through labelling and any other suitable means with a view to avoiding the occurrence of undesired allergic reactions. The possibility that individuals who are constantly exposed to insects, such as insects-workers, may develop also a related food allergy is an area that deserves further attention, especially in the case farming and production insect are scaled up in future (Ribeiro et al. 2018, 2019).

In the EU, EFSA, in its capacity as risk assessor, was questioned on the safety of insects for food and feed. Overall, it concluded that, while further research on the topic is needed, from a safety point of view, insects do not pose, in principle, any additional biological, chemical and environmental risk compared to other products of animal origin; nonetheless, substrates on which insects are fed and handling and storage of farmed insects may be pathways of contamination (EFSA 2015; Finkel et al. 2015; Paganizza 2016, 2019; Finardi and Derrien 2016; Grmelová and Sedmidusbsky 2017; Goumperis 2019; Monteiro et al. 2020). Similar conclusions, including the need to perform further research, have been reached by risk assessors of some EU Member States, including the Scientific Committee of the Federal Agency for the Safety of the Food Chain (AFSCA) in Belgium and the Agence nationale de sécurité sanitaire de l'alimentation, de l'environnement et du travail (ANSES) in France (see further Sect. 5.1.1.2).

Scientific research has likewise paid much attention to quality aspects of insects as food. In this context, nutrition value and health benefits of edible insects are probably those which have been investigated more in recent years.

The high number of species of edible insects that currently exist globally, associated with other determinants such their developmental stage and diet, makes particularly difficult to draw an average nutritional profile for this new potential food category (Belluco et al. 2013; van Huis 2013, 2015; Dobermann et al. 2017; Roos and van Huis 2017). However, in general terms, edible insects have been described in scientific literature as a valuable source of energy, proteins, fat, fibre and certain micronutrients (iron, zinc, calcium, potassium, magnesium and selenium, amongst others), with proteins and fats being the main components of insect composition (on average, 50–82% and 10–30% of dry weight, respectively) (Rumpold and Schlüter 2013). From this perspective, edible insects can be therefore compared to other foods of animal origin (e.g. meat, fish, eggs and milk) and play a role in a balanced and varied diet (Roos and van Huis 2017).

Until now the main focus of scientific research has been on edible insects as source of proteins. This has significantly contributed to the positioning of insects as suitable alternatives to meat in the context of the ongoing global transition towards more sustainable agri-food production systems and dietary patterns (Belluco et al. 2013). In this respect, current scientific knowledge indicates that insect protein fraction has great quality due to its content in essential amino acids and digestibility, being somewhat comparable to common protein sources. Conversely, the presence of chitin – a carbohydrate polymer present in the exoskeleton of arthropods and, thus, of insects too and an important source of fibre – adversely impacts protein digestibility and absorption. Removal of chitin may therefore guarantee the preservation of the quality of insect proteins (Belluco et al. 2013; Dobermann et al. 2017), while the use of chitin and its by-products could be studied in future for food supplementation as in the case of chitin extracted from shellfish (De Foliart 1992a). Notwithstanding the above, further scientific studies *in vivo* are needed to fully apprehend the inherent quality of insect proteins besides their digestibility in humans (Roos and van Huis 2017).

Furthermore, the role of edible insects as providers of high quantities of certain minerals, notably iron and zinc, should not be underestimated. These minerals, in fact, are essential to address specific nutrition deficiencies (e.g. anaemia and stunting) that are quite common in developing countries, especially amongst pregnant women, infants and children, mostly because of predominantly plant-based diets (Belluco et al. 2013; van Huis 2013; Roos and van Huis 2017). This considered, edible insects may well therefore be a suitable food category to be included in national food fortification programmes (van Huis 2020a).

In addition to providing essential nutrients, edible insects have been also identified as a source of various bioactive compounds. These are substances whose consumption can bring specific benefits to human health, in particular, by acting as risk-reduction factors of certain non-communicable diseases. By way of an example, several insect species are said to contain compounds with anti-oxidant properties, i.e. with the potential to reduce molecular damage in the human body, which are capable of inducing weight loss in obese subjects or which may reduce symptoms of health conditions like diabetes 2, hypertension and Parkinson's. However, most scientific findings in this area are the result of studies conducted on animals. This considered, in the EU, at least, identified health benefits will need to be further corroborated by human studies *in vivo* in order for food business operators of the insect sector to be able to lawfully make health claims on their products (Roos and van Huis 2017). From an industry perspective, appropriate scientific substantiation of any health and other claim that is made in relation to edible insects is perceived as fundamental with a view to establishing and maintaining consumer trust in the context of this emerging business sector (Aarts 2020).

2.6 Animal Welfare?

The prospect of the diffusion of mass-rearing of insects for food and feed purposes at global level does not raise only questions on how best to ensure safety and quality across all stages of their production, processing and distribution. As for other animals, ethical considerations essentially related to insect welfare in rearing establishments and at the time of slaughter may also come into play in this context. Even so, this is a scientific area to which, on balance, little attention has been paid to date compared to safety and quality aspects of insect farming (van Huis 2020c).

Nonetheless, looking ahead, insect welfare seems an area of no less importance policy wise, especially if one considers that animal welfare, in general, is usually regarded as a key component of the more virtuous agri-food systems, besides fitting well within the overall sustainability narrative subjacent to insects as food and feed (see above Sect. 2.4).

All in all, at present insect welfare raises some scientific questions whose solution would allow to clarify whether or not concerns over the well-being of these animals, when reared for food and feed purposes, are justifiable. In essence, up to now science has not been able to determine with enough certainty whether and to what extent insects are sentient beings, i.e. animals which are able to experience pain besides other emotional states (Lähteenmäki-Uutela and Grmelová 2016; Knutsson and Munthe 2017; Goumperis 2019; Kuljanic and Gregory-Manning 2020).

Several authors (for instance, Eisemann et al. 1984) have ruled out insects as sentient beings in the sense described above, admitting, instead, their capacity of experiencing 'nociception'. The latter term refers to the capacity of a being to perceive the imminence of risk and avoid it through a rapid response mechanism involving involuntary protective reflexes, though with no actual feeling of pain processed as such at the level of its brain (van Huis 2020c).

Other authors, including the above cited Knutsson and Munthen as well as van Huis et al. (2013), consider, instead, that, in the absence of scientific certainty on the matter at present, a cautionary approach is the preferred option for dealing with farmed insects. In essence, this approach implies that the suffering of those animals should be kept to a minimum. In this respect, it is relevant to note that, despite the absence of specific binding norms in this area even in countries where legislation is more mature, like in Switzerland (see further Sect. 4.5), some stakeholders, including the International Platform of Insects for Food and Feed (IPIFF) in Europe, are already advocating in favour of the adoption of good animal welfare practices by the business operators pertaining to the insect sector (IPIFF 2019a). From this standpoint, insect welfare may encompass, amongst others, minimum requirements for livestock population density in rearing establishments and killing techniques that minimise animal suffering (e.g. freeze-drying, heat and shredding). In particular, freezing is considered the most natural way to end insect life as it replicates the winter conditions in which those animals would normally die (Lamsal et al. 2019). Conversely, preventing cannibalism for those inspect species practicing it seems a more controversial aspect welfare wise, besides raising issues with regard to its practical implementation at the level of individual establishments.

2.7   Consumer Acceptance of Edible Insects

Whilst, in particular over the last decade, scientific research has significantly contributed to gaining a better of understanding of the benefits and the implications of incorporating insects in the human diet and in the current agri-food production system, acceptability of edible insects by consumers is a central pre-requisite for their establishment as food in the long run.

Currently, consumer acceptance of edible insects in Western societies faces a number of cultural and social obstacles (De Foliart 1999; van Huis et al. 2013; House 2017). In that context, entomophagy is often regarded as a primitive, animalistic or barbarian practice and traditionally associated with poverty, starvation or the occurrence of other exceptional circumstances, besides triggering mixed reactions amongst Western consumers (see above Sect. 2.2). Also, certain religious beliefs may play against the uptake of entomophagy in Western societies if one considers that most insects are not compatible with requirements of halal and kosher diets. Furthermore, as referred earlier on (Sect. 2.2), globally, survival of entomophagy is at risk in various countries whose traditional diets include insects as emulation or adaptation to Western lifestyles are major contributing factors for the progressive decline of insect consumption.

Nonetheless, the history of food shows us that cultural and social changes may happen under certain circumstances. For instance, both Johnson (2010) and van Huis (2013) point out to sushi and other typical Japanese raw fish preparations as one of the most recent examples of natural uptake of a non-traditional food in the majority of Western diets. Schelomi (2015) and other scholars consider the history of lobster as particularly relevant for the future acceptance of insects as food by consumers in Western societies. In fact, like insects, this crustacean is an arthropod too, which in the US passed from being considered a food unworthy of being eaten even by prisoners to be the expensive delicacy that we know today. This last example suggests, inter alia, that campaigns and raising-awareness actions aimed at promoting insect consumption could emphasise the biological relation between crustaceans and insects to win over part of the current consumer reluctance towards entomophagy.

Against this background, during the last decade, consumer field research has been conducted with a view to gathering information on psychological (e.g. food neophobia, disgust, aversion), cultural (e.g. tradition, religion), social (e.g. status, age, sex) and economic (e.g. affordability, price) determinants which may influence consumer behaviour towards edible insects. Most of this research – even if some scholars consider it still of limited scope (Schelomi 2015) – has focussed on Western markets, including the USA and the EU/Europe, where entomophagy, in spite of being still in its infancy, is considered to bear the greatest economic potential (Schlups and Brunner 2018).

In the EU, early consumer research carried out in this area and in several different countries essentially explored the overall acceptability by local consumers of edible insects, commonly known as 'willingness to consume' (WTC). For instance, with reference to the Czech Republic, Bednářová et al. (2013) concluded that, for their sensory qualities, larvae and brood of *Apis mellifera*, larvae of *T. molitor* and nymphs of *Gryllus assimilis* ranked best in terms of consumer satisfaction amongst a list of possible insect species suitable for human consumption. Verbeke (2015) found out that, in Belgium, a relatively limited percentage of the Flemish population (12.8% and 6.3% of male and female respondents, respectively) stated to be ready to integrate insects as meat substitutes in their daily diet, mainly motivated by the desire to embrace more sustainable consumption patterns. Hartmann and Siegrist (2016) found that, overall, German consumers were more inclined towards eating insects if the animals were not clearly visible or discernible and, thus, processed and incorporated in composite products. Conversely, consumer research conducted by Kostecka et al. (2017) in Poland revealed that half of the interviewees would not eat their favourite food if it were to include insects. More recently, a study based on a Finnish population sample including vegan, vegetarian and omnivore consumers singled out vegetarians as the consumer group with the most positive attitude towards the possibility of eating insects, in spite of their diet ruling out meat consumption as such, whereas for vegans this act would be fundamentally immoral and socially irresponsible (Elorinne et al. 2019). Mancini et al. (2019) have performed an extensive review of all studies carried out to better understand attitudes of European consumers towards edible insects and identified some of the most recurring limitations. These include the application of methodologies which are designed for non-novel foods, the lack of representativeness of the sample analysed and partial disregard for key determinants of consumers' choice (e.g. product availability and product promotion).

Overall, the studies conducted so far in the EU have shown a relatively low WTC vis-à-vis edible insects in the near future, with food disgust and food neophobia being the most common triggers for rejection by consumers (Cunha and Ribeiro 2019). In this context, youngsters, for being more prone to new food experiences and often more environmentally conscious than older generations, generally represent the population segment more likely to embrace first insects as meat alternatives or novel food ('*early adopters*') (Verbeke 2015; Caparros Megido et al. 2016; La Barbera et al. 2018; Schlups and Brunner 2018).

These findings are largely in line with consumer research conducted in other Western countries outside the EU, namely in Australia and in the USA (Wilkinson et al. 2018; Ruby and Rozin 2019), but also in Brazil (Cheung and Moraes 2016). However, in the Western context, Switzerland seems to be quite an exception as for Swiss consumers neophobia, in particular, would not constitute a barrier to the consumption of edible insects. For Schlups and Brunner (2018) this might have to do with the level of awareness which media and the political debate that preceded the adoption of national legislation authorising the production and the commercialisation of certain species of edible insects in the country generated amongst citizens (see further Sect. 4.5).

In any event, now that several insect products are on the market of some EU Member States (Sect. 5.1.1.2), research can and should primarily focus and assess actual consumer preferences, attitudes and trends (House 2017). In line with this approach, for instance, recent consumer research conducted on the Belgian market has shown that local consumers tend to eat insects, in particular, as ingredients of other food products (e.g. energy bars and shakes as well as burgers) with supermarkets being their preferred place for the purchase of those products (Van Thielen et al. 2019). If compared with earlier studies performed when edible insects were not offered for sale in Belgium and the Netherlands, the latest ones reveal positive signals in terms of willingness to consume (Mancini et al. 2019).

Conversely, in the EU the use of insects as animal feed intended for food-producing animals would appear, all in all, to be much less controversial from a consumer perspective. This because, on the one hand, insects are already part of the natural diet of several food-producing animals (van Huis 2013); on the other hand, the first studies in this area have shown that farmers, consumers and other stakeholders consider this practice acceptable to a large extent (Neves 2015; Verbeke et al. 2016; PROteINSECT 2016; Popoff et al. 2017). This may suggest that turning insects into common and ordinary feed could ultimately help pave way to their gradual acceptance by Western/European consumers as food too (Dobermann et al. 2017). In any event, more research is needed on the quality of food produced by animals other than fish which have been fed with insects (Gasco et al. 2019).

All this considered, it is legitimate to ask ourselves if edible insects do really stand a chance to become part of Western diets in future. In this respect, some commentators consider that up to now diffusion of edible insects in Western countries has clearly failed and this notwithstanding several decades of advocacy by entomologists and insect-eaters alike (Schelomi 2015).

As already mentioned above (Sects 2.3 and 2.4), over the last decade, the narrative developed by international organisations and stakeholders to promote the consumption of edible insects has essentially been centred on their potential in terms of food security and environmental sustainability as opposed to the production of conventional meat. On the one hand, this narrative has contributed to the identification of new business opportunities by food business operators as well as to the gradual structuring of an insect sector at global level. On the other hand, it has helped raising awareness and interest about edible insects, in the first place, amongst those consumers whose purchase decisions tend to be motivated primarily by environmental, societal and/or ethical values (e.g. millennials, vegetarians, flexitarians).

Considering the findings of consumer research in Western countries presented above, it is quite possible that this narrative alone will not be able to deliver the cultural change that the diffusion of edible insects requires (Deroy et al. 2015; van Huis 2017). Rather such arguments are considered by some commentators as capable of producing the opposite effect, i.e. intensifying aversion towards edible insects in Western consumers (Loo and Sellbach 2013).

In this respect, Schelomi (2015), by applying the theoretical model on the diffusion of innovation developed by Rogers (2003), considers that the uptake of entomophagy in Western societies essentially depends upon consumer acceptance of insects as a food innovation. Amongst others, this means that, in order for consumers to integrate insects in their diets, entomophagy should provide them with a relative advantage, which include social benefits (prestige or status), satisfaction (e.g. pleasurable appearance, texture, taste, flavour etc.) and convenience (e.g. economic affordability).

It is unquestionable that at present the conditions referred to above are hardly met by edible insects, which are not perceived by Western consumers as neither fancy nor luxury food and, notably as meat substitutes, have still a long way to go to be able to compete on an equal foot with meat products in terms of organoleptic characteristics and consumer prices (van Huis 2013, 2020a).

From this perspective, only the fulfilment of the conditions referred to above can contribute to transforming edible insects from a food novelty into mainstream food or food ingredients. In parallel, the design of supply chain strategies where insects are used to replace systematically other ingredients in staple foods, which guarantee a wide range of insect-based products and enhance their sensory properties may help drive and stimulate consumer demand of edible insects. All in all, the expectations of Western consumers need to be well reflected in market strategies of insect business operators for these to be persuasive and ultimately successful (Deroy et al. 2015; van Huis 2017, 2020a; Aarts 2020).

Chapter 3
The Global Market of Insects as Food and Feed

3.1 Characterisation of the Global Market

A characterisation of the global market for insect as food and feed is no easy job as comprehensive business studies are still lacking in light of the fact that the sector operates differently across the world and that, in certain geographical areas, edible insects, in particular, are a relatively recent novelty.

As far as insects intended for human consumption are concerned, a significant growth is forecast over the next decade with production volumes expected to reach globally over 730,000 t in 2030, corresponding to almost 8 billion US$. Protein bars and protein shakes are the insect-based products which are expected to undergo the highest growth, mainly because of the growing importance attached by younger generations to health and wellness. North America is the market with the highest predicted increase in consumer demand of insect-based products. Currently, out of the various insect species suitable for human consumption, crickets seem to dominate the market in terms of food applications (Meticulous Research 2019). Besides that, sales modalities may vary significantly depending on the country or world's regions. Overall, as Pippinato et al. 2020 report, these may consist of:

– short supply chains, which essentially rely on the performance of direct sales by insect farmers to final consumers;
– more complex commercial circuits involving a variable number of intermediate players and/or middlemen; or
– both of the above.

Regarding the use of insects for animal nutrition, i.e. pet food and feed, globally, the largest market in terms of production volumes and demand is currently Europe, followed immediately after by North America. However, in the medium term, Asia-Pacific is the region which is expected to record the highest growth in terms of demand of insect-based feed by 2030, because countries such as

F. Montanari et al., *Production and Commercialization of Insects as Food and Feed*, https://doi.org/10.1007/978-3-030-68406-8_3

China and India are gradually embracing the practice of feeding livestock with insects (Persistent Market Research 2019).

Against this background, Table 3.1 provides a list of 16 selected feed and food business operators of the insect sector based in the non-EU countries whose regulatory frameworks are analysed in Chap. 4. This list has been compiled with the objective to have a representative sample including market leaders from those countries, identified mainly through desk research and / or the websites of national trade associations, where the latter exist (e.g. USA, Canada, Australia). It should nevertheless be noted that, by any means, such a list cannot be considered as exhaustive. The high number of companies (amongst which several microbusinesses) located in the different countries and, ultimately, data availability make this a very arduous task at this point in time.

Based on the analysis of the data collected, while bearing in mind the inherent limitations of the studied sample, some observations can be made which may help better understand how the global market of insects as food and feed is currently developing.

Overall, the large majority of the companies studied in this sample (n = 11) operate in the insect food market segment. Yet, only two companies of the sample trade in both the food and feed market segments. Furthermore, four different species are used for the commercialisation of food applications with the house cricket (*A. domesticus*) being by far the most common. Conversely, a higher number of insect species is used for feed applications with the black soldier fly (*H.illucens*) and the house cricket being the most common ones which are reared and commercialised as feed.

In terms of business model, the large majority of the companies of the sample analysed (n = 14) have opted for integrated solutions covering rearing of insects for food or feed, their processing and distribution (*business-to-business*– B2B) or retail (*business-to-consumers*– B2C). It is also worth noting that almost all companies of the sample which operate in the insect food market segment (9 out 11) are currently selling their products to final consumers: protein bars and insect powder / meal are the most common products which they commercialise.

3.2   Characterisation of the EU Market

Of the different international markets where the insect food and feed industry is currently established and steadily developing, the EU market stands out in terms of number of companies, range of existing and potential applications and importance of financial investments supporting research and development activities. Likewise, investments aimed at scaling up insect production have been substantial: 600 million EUR in 2019 with a forecast of 2.5 billion EUR in 2020 (IPIFF 2020a). In addition to that, it should not be underestimated that the European insect sector can rely today on a trade association (IPIFF) which has considerably contributed to raising awareness about the objectives and the activities of the sector with policymakers and the general public across the EU.

Table 3.1 Overview of selected feed and food business operators of the insect sector in the non-EU countries covered in Chap. 4

Business operator	Market segment	Country	Business model				Insect species	Products
			Rearing	Processing	Distribution (B2B)	Retail (B2C)		
(Aketta/Bugs)	Food	USA	✓	✓		✓	A. domesticus	Cricket powder
CHAPUL	Food	USA	✓	✓		✓	G. sigillatus	Protein powders and bars
ASPIRE FOOD GROUP	Food	USA	✓	✓		✓	A. domesticus	Protein and energy bars
ENVIROFLIGHT	Feed	USA	✓	✓	✓		A. domesticus	Animal feed and fertiliser
AKETTA	Food	USA	✓	✓		✓	H. illucens	Protein bars and powder
essento	Food	Switzerland	✓	✓	✓	✓	A. domesticus	Protein bars and snacks
							T. molitor	
entomos	Food	Switzerland	✓	✓		✓	A. domesticus	Snack, bars, cookies, lollipops and bread
							L. migratoria	
insekterei	Food	Switzerland	✓	✓	✓		L. migratoria	Powder and biscuits
							T. molitor	
Enterra	Feed	Canada	✓	✓	✓		A. domesticus	Pet food and animal feed
							G. sigillatus	
							T. molitor	
oreka	Feed	Canada	✓	✓			H. illucens	Aquaculture feed
ENTOMO FARMS	Food & Feed	Canada	✓	✓		✓	H. illucens	Snacks, powders &pet food
SCHUBUGS	Feed	Australia	✓		✓		A. domesticus	Live insects
							A. domesticus	

(continued)

Table 3.1 (continued)

Business operator	Market segment	Country	Business model				Insect species	Products
			Rearing	Processing	Distribution (B2B)	Retail (B2C)		
REBEL Food Tasmania	Food	🇦🇺	✓	✓	✓	✓	*A. domesticus* *T. molitor*	Protein butters and live insects
GRUBSUP	Food	🇦🇺	✓	✓		✓	*G. sigillatus*	Bars and powders
NUTRINSECTA	Food & Feed	🇧🇷	✓	✓	✓		*T. molitor* *A. domesticus*	Animal feed and dehydrated insects
Salvi	Feed	🇧🇷	✓	✓	✓		*T. molitor* *G. assimilis* *Z. morio* *N. cinérea* *B. giganteus*	Pet food

Source: Montanari, Pinto de Moura and Cunha (2020) based on desk research

Other than that, because of the current regulatory framework in force in the EU (see further Chap. 5), it is clear that the use of insects for the purpose of animal nutrition is much more advanced and mature as opposed to insect food applications.

Against this background, the following sections provide a detailed characterisation of the European insect market for food and feed, respectively.

3.2.1   Food

The European market of edible insects is still in its infancy and, as such, has not been subject to in-depth business analyses as of yet. Nonetheless, policy and academic interest in this respect is growing and, as a result, few market studies have been recently conducted (IPIFF 2020b; Pippinato et al. 2020).

According to IPIFF (2020b), there are currently around 70 companies operating in the European market of insects for human consumption. Overall, the most numerous players in this market segment are micro-companies, i.e. businesses with less than 10 employees (81%). Conversely, medium-sized companies (50–250 employees) represent a very small portion of the total (3%). Financial investments supporting companies' activities are overall in line with this business landscape, with the majority of such investments (63%) being below 500,000 EUR. In terms of geographical distribution, research by Pippinato et al. (2020) shows that the highest concentration of insect food companies can be found in Northern Europe, with the United Kingdom, Germany and Belgium accounting for the largest share.

As any other food sector, the edible insect sector encompasses too different business activities, in line with the conventional 'farm-to-fork' approach. From this standpoint, research conducted by IPIFF (2020b) shows that, currently, on the European market, the majority of businesses of this market segment (over 1/3) deal exclusively with the processing of ingredients derived from insects, which are therefore intended to be incorporated in other food products. Another significant portion (28%) though is involved in all the stages of the production chain of edible insects, thus covering farming, processing and distribution or retail (B2B and/or B2C). Conversely, the number of companies whose business model only focuses on rearing insects for human consumption or on their sale at retail level is marginal (3% in both cases). In terms of species, the European insect food sector currently covers the following ones: the black soldier fly (*H. illucens*), the yellow mealworm (*T. molitor*), the lesser mealworm (*A. diaperinus*), the house cricket (*A. domesticus*), the banded cricket (*Gryllodes sigillatus*) and the migratory locust (*L. migratoria*).

Based on the same research, whole insects presently account for the highest share of the European market (over 1/5), followed by bars, biscuits and snacks that contain insect products (e.g. insect meal) as ingredients. According to Pippinato et al. 2020, whole insects, alongside insect meal, would be particularly popular amongst European businesses for requiring limited processing. Yet, market forecasts indicate that, in the medium term, food categories such as sports foods, supplements and dietetic foods may become market leaders in the European insect food market, due to the contribution by insects towards the nutritional composition of those products in terms of protein content.

Moreover, production wise, in 2019 only 500 t of insects and insect products were produced and commercialised as food and food ingredients in the EU. While this limited production output is mostly due to the inherent complexity of the current regulatory framework applicable to edible insects at EU level (see further Sects. 5.1.1.2–5.1.1.4), the gradual opening of the European market to such products could boost the production up to 260,000 t in 2030.

Other than that, at present edible insects – as such or as ingredients of other food products – are mostly offered for sale through companies' websites, online portals and during in-person events, such as fairs and conferences, while the mass-catering sector is the least exploited trade channel. Obviously, the above referred regulatory complexity which presently characterises the EU market makes so that the full potential of certain trade channels (e.g. offline / conventional retail) cannot yet be exploited to the same degree in all EU Member States. In any event, as of today, approximately 9 million consumers in Europe would have already tasted food consisting of or containing edible insects, while, by 2030, 390 million consumers are expected to have done so. From a consumer angle, research from Pippinato et al. (2020) points out to the existence of significant price differences across the range of insect products which are currently available on the European market. The formation of consumer prices appears to be influenced, above all, by the insect quantity present in the final food product (thus, with whole insects being the most expensive product category, on average, and protein bars the cheapest one, proportionally) and, secondly, by the insect species (e.g. production costs, availability, origin etc.).

Finally, in the medium term, the opening of the EU market to the production and commercialisation of certain species of edible insects is also expected to positively contribute to the employment market, namely with the creation of over 4,000 new direct and indirect jobs in Europe by 2030.

3.2.2 Feed

As referred to earlier on, the production and commercialisation of insects as animal feed is much more developed than their use as food. As it will be shown more in detail under Sect. 5.2, this situation is largely attributable to the existence in the EU of a more liberal legal framework for insects as feed as opposed to food. In particular, such a framework currently allows the use of insects as:

(i) pet food, including dead/live insects, insect fat and insect proteins; and
(ii) livestock feed: live insects, insect fat and insect proteins though, in the latter case, only for certain animal species.

This considered, currently, pet food and insect proteins for aquaculture feed are the two main market segments for European insect feed manufacturers.

In particular, following the granting of the EU authorisation permitting the use of insect proteins as feed for farmed fish in 2017, over 5,000 t of insects have been commercialised in the European market, with aquaculture absorbing half of the EU insect production which is presently destined to animal nutrition. As similar authorisations are being considered at EU level for other animal species, namely for poultry and swine livestock populations, the insect feed sector has the potential to grow further with a forecast production of over 2.7 million t by 2030 (IPIFF 2020a). The extension of the range of permitted substrates on which insects can be fed – so as to cover, amongst others, food waste, catering waste and slaughterhouse products – is also regarded as another step which could contribute to scaling up even further the insect feed sector in the EU.

In terms of species, the European insect feed sector currently covers the following ones: the black soldier fly (*H. illucens*), the common housefly (*M. domestica*), the yellow mealworm (*T. molitor*), the lesser mealworm (*A. diaperinus*), the house cricket (*A. domesticus*), the banded cricket (*G. sigillatus*) and the field cricket (*G. assimilis*) (IPIFF 2019b).

3.2.3   Market Leaders and Forerunners

For comparative purposes with the analysis performed in Sect. 3.1. (see Table 3.1) with regard to the global market, Table 3.2 provides for a list of 16 selected insect business operators based in the EU, which includes major market leaders for feed and key forerunners for food. Also in this case, the mapping of the actors realised cannot be considered as exhaustive, notably because of obvious language barriers and of the high number of companies present in the EU market, amongst which several microbusinesses and start-ups which are currently awaiting to step into that market.

In contrast with the situation described for the global market, a slight majority of the European insect business operators analysed operate in the feed market segment (n = 6), while less than 1/3 operates, respectively, in the food market segment or in both market segments. Furthermore, almost 2/3 of the companies analysed have opted for an integrated business model that combines rearing, processing and distribution (B2B) or retail (B2C) activities, although only four of them – not surprisingly all located in Northern Europe (see further Sect. 5.1.1.2) – are currently marketing their insect food products in the B2C market.

In terms of food applications, snacks, protein bars and insect powder / meal are the most common ones across the sample analysed. Amongst the insect species currently marketed or awaiting to enter the EU market, the yellow mealworm (*T. molitor*) is by far the most exploited, followed by the house cricket (*A. domesticus*) and the migratory locust (*L. migratoria*). In terms of feed, the black soldier fly (*H. illucens*) is the most used insect species within the sample considered.

Table 3.2 Overview of selected feed and food business operators of the insect sector in the EU

Business operator	Market segment	EU country	Business model				Insect species	Products
			Rearing	Processing	Distribution (B2B)	Retail (B2C)		
Ÿnsect	Feed		✓	✓	✓		*T. molitor*	Aquaculture and pet food, insect oil and fertiliser
micronutris	Food		✓	✓			*T. molitor** *G. sigillatus**	Whole insects, bars, snacks and pastas
nextProtein	Feed		✓	✓	✓		*H. illucens*	Dry protein powder, insect oil and fertiliser
innova feed	Feed		✓	✓	✓		*H. illucens*	Protein, insect oil and fertiliser
Next Earth	Food		✓	✓			*T. molitor**	Insect powder
HEXAFLY	Feed		✓	✓	✓		*H. illucens*	Pet food and animal feed
ILLUCENS	Feed		✓	✓	✓		*H. illucens*	Animal feed
MealFood	Food & Feed		✓	✓			*T. molitor*	Pet food, animal feed and fertiliser
PROTIX	Food & Feed		✓	✓		✓	*L. migratoria** *T. molitor** *A.domesticus**	Snacks, protein meal, insect oils, pet food, animal feed and fertilisers
Protifarm	Food & Feed		✓	✓		✓	*A.diaperinus** *L. migratoria* *T. molitor*	Snacks, protein meal, insect oils, pet food, animal feed and fertilisers
AGROLOOP	Feed		✓	✓	✓		*H. illucens*	Insect oil, protein meal and fertiliser

	Food/Feed					Species	Product
snack-insects.com	Food		✓	✓	✓	A. domesticus L. migratória T. molitor A. diaperinus	Snacks, pasta and chocolates
ENORM	Food& Feed		✓	✓		H. illucens*	Insect powder and animal feed
KRIKET	Food		✓	✓	✓	A. domesticus	Snacks and chocolates
ENTOMOBIO	Food		✓	✓		T. molitor	Insect powder and gin with insects
Italian Cricket farm	Food & Feed		✓	✓		T. molitor L. migratória A. domesticus Z. morio	Animal feed, pet food, cricket powder and whole insects

Source: Montanari, Pinto de Moura and Cunha (2020) based on desk research
*Pending novel food application (see Table 5.4 in Sect. 5.1.4)

Chapter 4
Insects as Food and Feed: Analysis of Regulatory Experiences in Selected Non-EU Countries

4.1 The International Context

On an international level Codex Alimentarius is the main standard-setting body for the production and the commercialisation of food and feed. Founded in 1963 upon joint initiative of two other international organisations, notably FAO and the World Health Organization (WHO), its main mission consists in the development of technical standards with the aim to facilitate international trade and promote safety and quality of food in the interest of consumers globally.

From a legal viewpoint, Codex standards are not binding on countries which are its members. However, in the context of the World Trade Organization (WTO) and namely of the Sanitary and Phytosanitary Agreement (SPS), adherence to Codex sanitary standards by member countries creates a legal presumption that the measures adopted at national level are consistent with international trade rules. As such, Codex standards can play a fundamental role in the resolution of international trade disputes concerning food and feed. In accordance with article 2 (2) of the SPS Agreement, the adoption of more stringent measures – that is more trade-restrictive – by a Codex member needs to comply with the principle of proportionality and be scientifically justified.

Currently, there are no specific Codex standards for insects for human and animal consumption (Lähteenmäki-Uutela et al. 2018). For the purpose of the Codex *acquis*, insects are generally considered as unavoidable food defects or impurities. For instance, Codex Standard 152-1985 requires wheat flour to be free of abnormal flavours, odours and living insects as well as filth (which includes impurities of animal origin, including dead insects) in amounts that may represent a hazard to human health. In spite of the lack of specific standards for edible insects, in particular, some of the existing Codex standards – for instance, in the area of food hygiene, meat hygiene and hygiene in transport of food in bulk and semi-packed – could be usefully applied to these animals as well (Halloran and Münke 2014).

The need to develop international standards for edible insects has been brought up only once at Codex level to date. In 2010, in the context of the regular Codex meetings for the Asian region, the People's Democratic Republic of Laos put forward a proposal to develop regional standards for the production and the commercialisation of edible crickets (*A. domesticus*) and derived products with a view to protecting consumer health and ensuring fair commercial practices (FAO/WHO Coordinating Committee for Asia 2010; Paganizza 2016). At that time, this proposal looked quite ambitious as most producing-countries in Southern-Eastern Asia did not have any national legislation governing the rearing and the commercialisation of edible insects nor data on their trade (Rumpold and Schlüter 2013). In fact, still today most of them do not have national technical provisions and/or standards (Reverberi 2017). According to Laos' proposal, standardisation in this area – covering production, safety, hygiene and labelling aspects, amongst others – would have ultimately contributed to increase confidence in the safety and the quality of exports from Asian producing-countries to the West where entomophagy was gaining some popularity (FAO/WHO Coordinating Committee for Asia 2010).

Despite the endorsement of other Asian countries (e.g. Thailand, Malaysia, Cambodia), the proposal in hand did not advance further, mainly because of the lack of data on international trade that would justify any specific standardisation work in this area (Van Huis et al. 2013; Paganizza 2016). Still, it cannot be excluded that Codex work in this area will resume at some stage in future as Laos followed up with its proposal calling for the creation of a dedicated working group on edible crickets (Reverberi 2017). In addition to that, a global coalition of trade associations representing insect businesses in Asia, Australia, the EU and North America has identified Codex Alimentarius as the most appropriate institutional setting which can guarantee harmonisation of rules for this new emerging sector and, in doing so, ensure that the latter further develops on a global scale (Payne et al. 2019; IPIFF 2020a).

4.2   United States of America

Crickets (*A. domesticus* and *G. sigillatus*) are the insects most commonly farmed for human consumption in the USA at present together with mealworms (Payne et al. 2019; see also Table 3.1 under Sect. 3.1). Specific references to edible insects in US federal legislation are pretty scarce, owing also to the absence of a dedicated legal framework for novel foods as it exists in other countries (European Parliament 2015).

Pursuant to the Food Defect Levels Handbook, developed by the Food and Drug Administration (FDA) within the powers conferred to it by Section 110.110, Title 21 'Food and Drugs', of the Code of Federal Regulations (CFR) (FDA 1995; US Government 1996), insects are treated and regulated, first of all, as any other natural and unavoidable extraneous materials (e.g. mould, excreta by rodents and birds) that may negatively impact the aesthetic quality of food and whose presence in food may be the result of inadequate conditions or practices adopted during its production, storage, or distribution (Le Beau 2015; Boyd 2017; Gaulkin 2020). To this effect, the FDA Handbook above referred sets out, for a wide range of different food products or food categories, maximum limits which, if exceeded, render the food in question 'adulterated' and, as such, subject to enforcement in accordance with US food law. By way of an example, in frozen broccoli the presence of certain insects (aphids and thrips) and mites is no longer tolerated when their average number is 60 or more per 100 g. Yet, the legal framework set out by the Food Defect Levels Handbook does not apply to insects when these are purposefully used as or added to food.

Other than that, until recently the legal status of edible insects under US federal food law has been to some extent uncertain (van Huis et al. 2013; Lotta 2019) with the FDA being frequently singled out for its regulatory inaction (Le Beau 2015; Boyd 2017). Effectively, it is only as of 2015 that the FDA has started providing its own interpretation of the matter in public events and communications without, however, never formalising it in a proper guidance document or interpretative note (Boyd 2017; Watson 2017; Gaulkin 2020).

In any event, in accordance with FDA's current interpretation (FDA 2015), whole insects and insects minimally processed (e.g. milled) are to be considered as 'food' in accordance with US federal food law, notably pursuant to Section 201, lett. f), of the 1938 Food, Drug and Cosmetics Act. This means that business operators rearing insects, in particular, are subject to the same set of general rules that apply to any food operator. This includes, amongst others, the requirement that food must be clean and wholesome, that its production, packaging, storage and transportation must take place under adequate hygiene conditions and be properly labelled. Besides that, insects intended for human consumption must originate only from farms and production sites operating in line with good manufacturing practices – as wild insects may harbour a number of risks for consumers' health (e.g. diseases and contaminants) – while insects reared for animal nutrition cannot be diverted into the food chain.

Conversely, substances and products derived from insects, notably insect proteins, fall under the regulatory definition of 'food additives', pursuant to Section 570.3, lett. e), CFR: «all substances [...] the intended use of which results or may reasonably be expected to result, directly or indirectly, either in their becoming a component of food or otherwise affecting the characteristics of food». This definition is mirrored in the Food, Drug and Cosmetics Act (Section 321, lett. s) (US Government 1938). This interpretation would be confirmed by the fact that other proteins extracted from animals (e.g. fish) have been handled in the past as food additives by the FDA.

This considered, in accordance with Section 348 of the Food, Drug and Cosmetics Act, food additives are, as a rule, subject to pre-market approval, for which the same FDA is the responsible competent authority. Any person may, with respect to any intended use of a food additive, submit a petition for the adoption of a regulation laying down the conditions under which an additive can be safely used.

Alternatively, if the use of a given substance is 'Generally Recognized As Safe' (GRAS), in principle, there would be no need to undergo any pre-market approval. In accordance with US food law, GRAS status can be determined by the food business operator itself and solely under its own responsibility, although such a self-determination can always be challenged by the FDA. While the formal recognition of GRAS status by the FDA remains voluntary, it is increasingly required in B2B relations and can be a rather expensive procedure because of the scientific evidence which needs to be gathered to that effect. Another pathway to obtain GRAS recognition is to prove the safety of a substance based on its common use in food, although this option applies only to substances used in such a way prior to 1 January 1958 and is rarely chosen by business operators (Halloran and Münke 2014; Lähteenmäki-Uutela et al. 2017). By August 2020 no request for the recognition of GRAS status had been submitted to the FDA with regard to substances derived from insects.

In spite of the interpretation given by the FDA with regard to the legal framework applicable to insects intended for human consumption described above, doubts persist as to whether FDA's 'informal' guidance provides, in fact, enough legal certainty to the businesses of the insect sector to safely operate on the national market. While some would welcome a firmer regulatory action from FDA in this area, in particular by formally recognising insects as actual food instead of mere food defects, other analysts are of the view that ongoing regulatory developments in other world's regions – namely in the EU – may end up with exerting the right amount of political pressure on FDA to design a proper legal framework for edible insects (Gaulkin 2020). All in all, this would be quite an interesting and unusual spill over effect from the standpoint of international food policy if one considers that, generally, the USA has a more liberal approach than the EU when it comes to novel foods and biotechnologies, just to name a few.

As the notion of 'food' under US food law encompasses also animal feed and pet food, this makes so that insects for human and animal consumption are fundamentally subject to the same regulatory requirements.

4.3   Canada

Amongst the Western countries where entomophagy is not a traditionally rooted practice, Canada is possibly one of the most liberal ones in terms of regulatory approach vis-à-vis production and commercialisation of insects for human consumption. This is in part due to the rapid growth that the Canadian population has undergone over the last few decades, which is largely the result of immigration from developing countries and of the emergence of new consumers' preferences (e.g. ethnic food) (Government of Canada 2009). The first national rules regulating insects for human consumption were passed in mid-90s and concerned, in particular, the harvesting, hygiene, marketing and labelling conditions under which the larvae of *Gusano roja* could be used in the preparation of certain Mexican alcoholic beverages such as mescal and tequila (Halloran and Münke 2014).

As a federal State, Canada regulates the production and the commercialisation of insects as food and feed through laws and norms stemming different geographical and administrative levels (namely federal, provincial and municipal) (Halloran and Münke 2014; Lähteenmäki-Uutela et al. 2017).

As far as insects intended for human consumption are concerned, their permissibility is regulated at federal level. In accordance with Section 2 of the Canadian 1985 Food and Drug Act, insects are generally regarded as food («*food includes any article manufactured, sold or represented for use as food or drink for human beings, chewing gum, and any ingredient that may be mixed with food for any purpose whatever*») (Government of Canada 1985a, b). This interpretation of Canadian law essentially results from the assumption that several insect species have been part of the regular diet of the population of other countries for centuries (Halloran and Münke 2014). This considered, insects as food must fulfil the basic requirements that the national 1985 Food and Drug Act sets out in Sections 4, 5 and 7 for any other food, produced in the country or imported, that is safety, suitability for human consumption and fair practices vis-à-vis consumers.

Notwithstanding the above, in some cases insects may qualify for as novel foods under Division 28 of the Food and Drug Regulations (Government of Canada 2019). This provision stipulates that novel food, inter alia, means «*a substance, including a microorganism, that does not have a history of safe use as a food*». This means that they require a pre-market notification to the Food Directorate of Health Canada, the federal competent authority, proving their safety before they can be safely and lawfully placed on the national market (Halloran and Münke 2014; Halloran et al. 2015). Crickets, for instance, which are farmed in the country for human consumption (see also Table 3.1 under Sect. 3.1), are not considered novel foods in Canada (Reverberi 2017).

If insects qualify for as food or once they are authorised as novel foods, provincial laws may then still set out specific conditions for their production and commercialisation, including by mass-caterers, which may limit or even prohibit those activities in the relevant jurisdiction (Halloran and Münke 2014).

Currently, there are no specific conditions for the labelling of insects destined to human consumption, although most products on the Canadian market display a warning alerting people with allergies to crustaceans that they may suffer undesired reactions from their consumption (Lähteenmäki-Uutela et al. 2017).

In Canada, feed is regulated through federal laws, namely by the Feeds Act and the Feeds Regulation (Government of Canada 1985a, b and 1983). Insects without a history of safe use in animal nutrition are generally regarded as novel feed or feed ingredients and, as such, are subject to a prior safety evaluation by the Canadian Food Inspection Agency (CFIA). Authorisation dossiers submitted by novel feed applicants must contain relevant information, including, amongst others, a full characterisation of the insect species, production process, hazard identification, quality control and results of efficacy trials of feed for each targeted animal species. In principle, the authorisation procedure should have a duration of 90 days; nevertheless, regulatory practice shows that this time-limit is often exceeded in the case of insects, mainly owing to their relatively new status as feed at country level and lack of data (Lähteenmäki-Uutela et al. 2017). Detailed regulatory requirements applying to insect-based feed ingredients are currently being considered by CFIA following a consultation that took place in 2019 (CFIA 2019). To date CFIA granted only a few authorisations regarding insects as animal feed. These include the approval of the whole dried black soldier fly larvae (*H. illucens*) as livestock feed for broilers, salmonids and tilapia (Lähteenmäki-Uutela et al. 2017, 2018; Koeleman 2017).

4.4   Australia

Amongst the 'Western' countries that are analysed in this Chapter, Australia stands out as one of the most conservative when it comes to entomophagy and this in spite of this practice being part of the culture of the indigenous populations inhabiting the continent.

On the one hand, aversion by Australian consumers towards eating insects has somehow prevented the edible insects' industry from scaling up, whilst, over the last few decades, indigenous populations have been increasingly embracing Western diets. On the other hand, rearing of native insect species for human consumption has proven to be not always commercially viable (some of the native species take up to 5 years to reach an edible size), whereas the import of non-native species for the same purposes raises, in general, serious environmental and bio-security concerns (Yen 2010; Bartrim 2017). This has led in turn to lower public and private investments in research and development compared to other areas of the world (notably, Europe) (Bartrim 2017).

Notwithstanding the above, today the national industry recognises the potential of insects as alternative protein sources with reduced environmental impact. This has eventually resulted in the establishment of a national trade association, known as the Insect Protein Association of Australia (IPAA), which presently counts seven companies as members.

For the reasons illustrated above, the regulatory framework for insects intended for human consumption in Australia is not yet fully developed. Overall, under the Australian legal order edible insects may fall under three different product categories (Lähteenmäki-Uutela et al. 2017):

1. novel foods requiring pre-market authorisation;
2. traditional foods;
3. non-traditional foods with no safety implications.

To date the national scientific risk assessor, the Food Standards Australia and New Zealand (FSANZ), has recognised three insect species – *A. domesticus*, *T. molitor* and *Zophobas morio* – as non-traditional foods with no safety issues and, thus, as suitable for human consumption. In 2019, however, the same body supplemented its scientific opinion stating that «[*t*]*here is recent evidence to suggest there may be a risk of allergenicity in crustacean-allergic or other sensitive individuals when consuming crickets or foods derived from crickets*» (FSANZ 2020).

Other than that, insect farms are subject to prior authorisation by the relevant local competent authorities for biosecurity reasons and, when approved, must observe general good rearing practices. Additionally, food businesses processing insects must comply with general hygiene requirements set out in Section 3.2.3 of the Australia & New Zealand Food Standard Code. Lastly, businesses importing insects from other countries must make sure that animals are not endangered species protected under international law and that, when introduced on the Australian territory, they are dead and previously heat-treated (IPAA 2020a).

In Australia, insects can also be reared, processed and commercialised as animal feed. Unlike feed additives, supplements and veterinary medicines, feed ingredients and feed materials, including those containing or consisting of insects, do not require pre-market approval at country level (Lähteenmäki-Uutela et al. 2017). According to the national insect protein association, there would be currently 45 individuals and feed business operators rearing insects for animal nutrition in the country (IPAA 2020b).

4.5   Switzerland

Insects intended for human consumption can already be found on the Swiss market since 2017. Effectively, the Swiss legislator took advantage of a major review of the domestic food law system, amongst others, to ensure alignment with the new rules introduced by the EU for novel foods under Regulation (EU) 2015/2283 (see further Sect. 5.1.1.3).

Beforehand, whilst non-commercial breeding of insects and consumption of wild insects had been generally allowed in Switzerland, the production and commercialisation of insects for commercial purposes was not. Changes to the national food law system in this respect are the result of raising-awareness sustainability campaigns led by national NGOs (e.g. Grimiam), which eventually drew the attention of national politicians to the importance of entomophagy (Halloran et al. 2015).

For the purpose of the new legal framework in place in Switzerland since 1 May 2017, insects are generally considered as *'nouvelles sortes de denrées alimentaires'* – that is novel foods – within the meaning of article 15 (1) lett. e) of the Ordonnance sur les denrées alimentaires et les objets usuels du 16 décembre 2016. This provision classifies as novel any *«foods consisting of animals or their parts or which are isolated or produced from animals and their parts, with the exception of those foods derived from animals reared in accordance with traditional rearing practices before 15 May 1997 as long as they have a history of safe consumption as food in Switzerland»* (Conseil fédéral suisse 2016).

As a rule, novel foods such as insects are subject to prior authorisation in accordance with the procedure set out in article 2 of the Ordonnance sur les nouvelles sortes de denrées alimentaires du 16 décembre 2016 (Département fédéral de l'intérieur 2016a). This involves the submission of an authorisation request to the Office fédéral de la sécurité alimentaire et des affaires vétérinaires (OSAV) containing, amongst others, a description of the product, its composition and technical specifications, production method, multiplication or reproduction practices, data proving its safety and the lack of misleadingness, conditions of use, where applicable, and labelling requirements. Whenever the authorisation request is deemed to fulfil the requirements set by Swiss law to this effect, the applicant is then entitled to place the novel food at issue on the market for 5 years with no possibility of renewal. After such a period, insofar as the novel food still meets the conditions for being placed on the market, the exclusivity of which the applicant has benefitted ceases to exist and all business operators can commercialise it, in accordance with article 17 (2) of the Ordonnance sur les denrées alimentaires et les objets usuels du 16 décembre 2016.

Notwithstanding the above, Swiss law provides for the possibility of certain novel foods to be placed on the national market without prior authorisation. This scenario must be though considered as exceptional and, in this spirit, the Swiss legislator has made use of this derogation in a quite limited number of cases since the entry into force of the new national food law system. This includes the authorisation granted *ex lege*, since 1 May 2017, to the production and commercialisation of few insect species for human consumption, notably, yellow mealworm (*T. molitor*), house cricket (*A. domesticus*) and migratory locust (*L. migratoria*), pursuant to article 6 (1) lett. a) and under the conditions Annex 1 of the Ordonnance sur les nouvelles sortes de denrées alimentaires du 16 décembre 2016. This authorisation is generic (i.e. not associated with a specific holder) and is granted for the use of the insect species referred to above as food or food ingredients. Yet, it does not extend to proteins extracted from those insects nor cover other insect species, which should be all subject to *ad hoc* authorisations (Office fédéral de la sécurité alimentaire et des affaires vétérinaires 2017; Goumperis 2019). Table 4.1 lists the requirements set by Swiss law for the production and the commercialisation of insect species for which prior authorisation is not required.

National provisions allowing production and commercialisation of the insect species referred to above have been supplemented by guidance developed by the Swiss competent authorities (Office fédéral de la sécurité alimentaire et des affaires vétérinaires 2017). The guidance is a useful tool for food business operators willing to step into the insect market, insofar as it lists all relevant legal acts that must be complied with when rearing and commercialising insects in Switzerland.

The guidance clarifies that insect farms need registration with the cantonal competent authorities, while activities other than primary production, including processing and distribution, are subject to the granting of an administrative authorisation. Irrespective of the specific activity carried out, all business operators of the insect food chain are required to carry out their own controls to guarantee the applicable standards of food safety are complied with.

Besides that, insect farms are subject to general hygiene requirements, traceability rules and obligations to ensure that different production lines are duly segregated (e.g. rearing of insects intended for human consumption takes place separately from farming of insects for feed use). From a primary production standpoint, insects are considered as farmed animals and, as such, generally subject to national legislation

Table 4.1 Insect species that can be produced and marketed in Switzerland since 1 May 2017 with no prior authorisation, under the conditions set out in Annexe 1 of the Ordonnance sur les nouvelles sortes de denrées alimentaires du 16 décembre 2016

Insect species	Development stage	Legal requirements
T. molitor	Larvae	*Primary production:* – Insects must originate from a farm
A. domesticus	Adult	*Commercialisation:* – Before their placing on the market, insects must be quick-frozen for an appropriate period of time and subject to heat treatment or any other suitable process to eliminate vegetative germs – Insects can be commercialised in the following forms: whole, crushed and ground
L. migratoria	Adult	*Presentation and labelling:* – The name of whole insects must contain the name of the animal species (customary name + scientific name) – When insects are used as an ingredient in another food, the name of the latter must contain a reference to the insect – General labelling requirements set by national law for ingredients that may provoke allergies or other undesirable reactions apply by analogy (art. 11 of the Ordonnance concernant l'information sur les denrées alimentaires du 16 décembre 2016) (Département fédéral de l'intérieur (2016b)

applying to them (e.g. feed, veterinary medicines). For the killing of insects, the guidelines in hand recommend resorting to treatments relying on cold (e.g. freezing, quick-freezing) or CO_2 as they generally ensure that insects preserve their integrity. Conversely, no specific animal welfare standards are currently set out for insect farms.

As regards food business operators who perform further processing of insects and insect-based food products, these must apply the specific treatments required by Swiss law before commercialisation takes place. Besides that, they are bound to have appropriate systems in place to monitor hazards and critical points of the relevant production process. Neither specific microbiological criteria nor maximum levels for contaminants are currently set at national level for the insect species authorised to be commercialised as food. However, certain microbiological criteria applying to all foods (e.g. *Salmonella*, *Listeria monocytogenes)* can be applied to the processing of insects for human consumption.

Regarding specific aspects of the labelling of insects and insect-based products, the guidelines recommend the inclusion of a specific warning on the packaging to alert consumers that insect consumption may trigger allergic reactions in those suffering from allergies to molluscs and crustaceans and/or to acarian. In addition to that, the labelling of insects commercialised as whole may contain, where appropriate, an indication that legs and wings need to be removed before eating.

It is worth noting that since the novel food authorisation of the three first insect species which was granted in 2017, no new applications were submitted for the assessment by the Swiss authorities (Office fédéral de la sécurité alimentaire et des affaires vétérinaires (2020). Overall, the lack of applications observed to date may be due to strategic considerations by businesses operating in the insect market. Besides the fact that Switzerland is a smaller market as opposed to the neighbouring EU market, novel food authorisations granted by the EU guarantee automatic access to the Swiss market (i.e. no authorisation is needed to that effect), but the reverse does not apply (Département fédéral de l'intérieur 2016a). In any event, in 2019 Switzerland together with Canada was one of the first countries to be granted a general authorisation to export to the EU live and dead insects and products thereof (see further Sect. 5.1.2.1).

4.6   Brazil

In Brazil there is a scarcity of information and absence of official regulations or guidelines about insects as food and feed. Currently, on the national market insects are mostly used as pet food. Two different competent authorities are responsible for supervising this business sector: the Ministry of Agriculture, Livestock and Supply (MAPA), which controls products of animal origin, whether for human or animal consumption, and the National Health Surveillance Agency (Anvisa), linked to the Ministry of Health, which is responsible for sanitary controls of the production and commercialisation of food products.

At present, in the country the consumption of insects as food is possible only in two situations (Coutinho 2017):

– as a gastronomic experience, i.e. an experience which is occasional and, in any event, is not performed in a regular way; or
– for the purpose of preserving indigenous culture.

This considered, the few insect producers who are based in Brazil, especially those operating in the food business segment, target export markets or reportedly work through informal trading channels.

Currently, the only legislative reference related to insects can be found in RDC No 14/2014, which contains the tolerable limits for the presence of foreign matter in food (Anvisa 2014). This legislation considers insects, inter alia, as *«foreign matter indicative of risks to human health»* (e.g. cockroaches) and sets in Annex I the tolerance limits for the presence of insect fragments resulting from inadequate practices in various food categories.

With regard to insects as food, recently, Anvisa, through its press office, has clarified that companies interested in commercialising food products containing or derived from insects in the country must introduce a request for an authorisation pursuant to national legislation on novel foods. According to RDC No 16/1999, novel food and /or novel ingredient are *«foods or substances with no history of consumption in the country, or foods with substances already consumed, but which are added or used at much higher levels than those currently observed in foods used which are part of the regular diet»* (Anvisa 1999a). As a rule, novel foods are subject to a pre-market risk assessment by Anvisa based on RDC No 17/1999 (Anvisa 1999b). In general, Anvisa's risk assessment is based on the analysis of the information provided by the applicant regarding the purpose and conditions of use of the food and the evaluation of the risk assessment performed by the applicant itself.

The national novel food legislation is currently in the process of being reviewed and, in this context, the specific situation of insects as food will be most likely considered and discussed.

Until now, there are no official records of any novel food application for food or ingredients containing or derived from insects. However, it will not be long before the first application appears, as interest by food business operators in this area has been steadily increasing over the last few years. One proof of the growing business interest towards insects as food is the recent establishment of a national association of insect producers, Associação Brasileira dos Criadores de Insetos (ASBRACI), whose main objectives are to represent the interests of the sector before the competent authorities and contribute to its professionalisation and regulation.

As regards rearing of insects, there is also no specific regulation in the country. However, MAPA has started allowing the use of insects for animal consumption. Recently, this government body has licensed the first insect company under national legislation governing the manufacturing of ingredients for animal feed. Thus, MAPA may authorise breeding sites in the country, which, in order to operate, must hold a seal known as SIF (Federal Inspection Service), which ensures the quality of edible and non-edible products of animal origin.

Chapter 5
The EU Regulatory Framework for Insects as Food and Feed and Its Current Constraints

5.1 Food

There is no question that the potential of the production and commercialisation of edible insects in terms of food security and sustainability (see Sects. 2.3 and 2.4) has produced a considerable impact on the EU market and its regulatory framework over the last decade.

But can insects be lawfully destined to human consumption at all in the EU?

In order to answer this question, consideration should be given, in the first place, to the definition of 'food' pursuant to EU law. The latter is currently enshrined in Regulation (EC) No 178/2002, otherwise known as 'General Food Law', that is the overarching EU legal text governing the production, processing and distribution of food and feed (European Parliament and Council 2002; Paganizza 2016). In accordance with article 2 of that regulation, food is «*any substance or product, whether processed, partially processed or unprocessed, intended to be, or reasonably expected to be ingested by humans. "Food" includes drink, chewing gum and any substance, including water, intentionally incorporated into the food during its manufacture, preparation or treatment [...] "Food" shall not include: [...](b) live animals unless they are prepared for placing on the market for human consumption; [...](h) residues and contaminants*».

The wording of this definition provides the interpreter with some arguments to make a convincing case for framing insects as actual food under EU law. In the first place, in spite of the fact that insects are currently not part of the conventional diet of the vast majority of European population, it has been shown that entomophagy has been practiced locally in the European region to some extent (see Sect. 2.2). Furthermore, the presence of large communities of immigrants in the EU from geographical areas where insects are traditionally eaten may contribute towards the demand for edible insects, as long as the latter is properly encouraged or, in any event, not disincentivised (Paganizza 2016). Also, the very notion of what can be

F. Montanari et al., *Production and Commercialization of Insects as Food and Feed*, https://doi.org/10.1007/978-3-030-68406-8_5

reasonably expected to be ingested by a human being is subject to significant variations across the EU (such as in the case of horsemeat as referred by Grabowski et al. 2013). On this basis, one could argue that, all in all, insects, whether processed or unprocessed, maybe considered as products which can be «*reasonably expected to be ingested by humans*». This would justify, in turn, their deliberate use as food ingredients in the manufacturing and processing of other compound food products.

The reference contained in the definition under exam to the '*intentional*' incorporation of a food as an ingredient into another food product is of particular relevance in the case of insects. Effectively, in accordance with the EU administrative practice in the area of official controls, such animals have been generally considered as foreign bodies, i.e. extraneous matters that may involve a physical risk to human health, on a par of dust and plastic, metal or glass fragments and whose presence in a food, similarly to residues, contaminants and any other undesirable substances, is regulated, in principle, as adventitious (Belluco et al. 2013). Accordingly, in 2018 several instances of presence of live or dead insects and/or of insect excrements were notified by EU Member States through the Rapid Alert System for Feed and Food (RASFF), the majority of which concerned cereals, bakery products and fruit and vegetables (European Commission 2018).

Finally, live insects may fall into the EU definition of food insofar as they are prepared and destined to human consumption, as is the case for few other categories of animals (e.g. oysters).

Other than that, specific references to insects intended for human consumption were pretty scanty in EU legislation before the most recent developments which are further described in Sect. 5.1. In fact, the only explicit reference to insects as 'domesticated terrestrial animals' which can be exploited as 'livestock' could be found in article 2 of Regulation (EC) No 834/2007 on organic production and labelling of organic products (Council 2007; Paganizza 2016; Lotta 2019). Yet, the fact that insects are livestock in the sense of farmed animals for the purpose and in light of the scope of EU organic legislation, which is broader than food, cannot be reasonably used as additional evidence to conclude that insects are to be considered as food pursuant to EU law (Paganizza 2016).

5.1.1 Insects as Novel Foods

If, on the one hand, the fact that EU law potentially allows for the framing of insects as food has not been heatedly disputed in literature as well as at policy level, on the other hand, the identification of the specific regulatory regime applicable to edible insects has given rise to conflicting interpretations and solutions across EU Member States.

In this context, edible insects have been usually considered as products falling under EU legislation governing novel foods, notably under Regulation (EC) No 258/97 and, more recently, under Regulation (EU) 2015/2283, which replaced the former (respectively, European Parliament and Council 1997 and 2015).

As it will be shown in the present section, while some light has been eventually shed on the legal status of edible insects in the EU, some divergences in national approaches still persist today, preventing business operators of the food insect sector from operating on a level-playing field.

5.1.1.1 Regulation (EC) No 258/97

At the time of its adoption, in 1997, Regulation (EC) No 258/97 represented the very first legal framework introduced by the EU (then European Community) to regulate the placing on the market of novel foods and novel food ingredients. Overall, the existence of different national laws at Member State level in this area and the obstacles to the free movement of foodstuffs stemming from the lack of harmonisation motivated the adoption of a common legal framework at EU level, in addition to the need to guarantee an adequate level of protection of the health and the economic interests of European consumers.

In accordance with Regulation (EC) No 258/97, the classification of a food or a food ingredient as novel under EU law essentially depended on the fulfilment of two conditions, i.e. the fact that the food or the food ingredient:

(i) had not been used for human consumption to a significant degree in the EU (then Community) before 15 May 1997 (which corresponds to the date of the entry into force of the regulation) (*temporal condition*), where the reference to 'EU/Community' has to be interpreted as to any Member State, in accordance with the jurisprudence of the European Court of Justice (ECJ) in the case *HLH Warentebries Gmbh and Orthica BV* (European Court of Justice 2005);

(ii) fell into one of the categories listed by the regulation (*substantive condition*), which are reproduced in Table 5.1.

As soon as economic and scientific interest around edible insects sprung up in the EU, their possible regulatory status as novel foods and food ingredients started to be considered and debated by stakeholders and competent authorities alike.

Table 5.1 Categories of foods and food ingredients which could qualify for as novel foods under Regulation (EC) No 258/97

Article 1 (2) regulation (EC) No 258/97 – foods and food ingredients
(a) Containing or consisting of genetically modified organisms;
(b) Produced from, but not containing, genetically modified organisms;
(c) With a new or intentionally modified primary molecular structure;
(d) Consisting of or isolated from micro-organisms, fungi or algae;
(e) Consisting of or isolated from plants and food ingredients isolated from animals, except for foods and food ingredients obtained by traditional propagating or breeding practices and having a history of safe food use;
(f) To which has been applied a production process not currently used, where that process gives rise to significant changes in the composition or structure of the foods or food ingredients which affect their nutritional value, metabolism or level of undesirable substances.

Indeed, on the one hand, for most insect species no human consumption to a significant degree in the EU before 1997 could be claimed and demonstrated; on the other hand, certain insect products (e.g. proteins and extracts derived from the whole animal or, according to some, their parts such as head, wings, thighs etc.) could possibly fall under the category *'food ingredients isolated from animals'* in accordance with article 1 (2) (e) of Regulation (EC) No 258/97 (Belluco et al. 2017; Goumperis 2019; IPIFF 2019c).

Notwithstanding the above, great uncertainty revolved around the regulatory classification of whole insects as possible novel foods and food ingredients, in particular, as they could hardly fit into the notion of ingredients which are isolated from animals, in line with the textual wording of Regulation (EC) No 258/97 (Laurenza and Carreño 2015; Paganizza 2016; Gremelová and Sedmidubsky 2017; Lähteenmäki-Uutela et al. 2017; Rusconi and Romani 2018; Lotta 2019; IPIFF 2019b).

It should be noted that the classification of whole insects as novel foods and food ingredients under Regulation (EC) No 258/97 was far from being a merely theoretical issue, but had (and still has) important regulatory implications, if one considers that from a legal point of view (Paganizza 2016; Belluco et al. 2017):

– the placing on the EU market of novel foods and food ingredients is only possible after an authorisation procedure proving their safety for human consumption;
– foods and foods ingredients that are not deemed as novel, in the absence of EU provisions detailing requirements on their production and commercialisation, may be subject to national rules adopted by Member States and, when complying with such rules, can then freely circulate within the EU market by virtue of the principle of mutual recognition, unless specific safety concerns are raised by the competent authorities of another Member State.

This being said, the regulatory uncertainty around whole insects as novel foods / ingredients, alongside with the absence of a firm and clear official guidance on the matter from the European Commission, gave rise to divergent legal interpretations as well as policy and enforcement approaches amongst EU Member States (see further Sect. 5.1.1.2).

In this respect, it is also worth noting that, despite the fact that the new EU novel food framework currently in force eventually clarified the status of whole insects under EU law (see further Sect. 5.1.1.3), the issue is still being considered, analysed and debated by legal scholars and in the context of court cases. From this perspective, the case *Entoma SAS* (C-526/19), which the ECJ decided on 1 October 2020, bears great relevance for the ultimate definition of the regulatory controversy that surrounds whole insects as possible novel foods and food ingredients under Regulation (EC) No 258/97 (European Court of Justice 2019, 2020a, b). Table 5.2 provides a summary of the main proceedings of the case.

Table 5.2 Summary of Entoma SAS case (C-526/2019)

Facts
Entoma SAS is a French food company which markets whole insects for human consumption, amongst which locusts, mealworms and crickets. In 2016 the national competent authorities (notably, the Prefecture de Paris) ordered the suspension of the marketing of whole insects by Entoma SAS together with the withdrawal of the company's products present on the market, alleging that, in the absence of an authorisation proving their safety, they could not be offered for sale. From a legal perspective, the French competent authorities considered that whole insects *did* fall into the scope of Regulation (EC) No 258/97, i.e. the novel food legal framework which applied at the time of the case. Conversely, the company took the opposite view that whole insects consumed in their own right were *not* subject to the EU novel food regime then in force – which would be confirmed by the specific transitional measures subsequently introduced by Regulation (EU) 2015/2283 – and, as a result, their production and commercialisation did not require any prior authorisation proving their safety as food.

Legal proceedings
Entoma SAS challenged the decision of the French competent authorities until the matter was eventually referred to the Conseil d'État, the highest national administrative court of the French judicial system. With a view to obtaining an authoritative interpretation of the provision of EU law being disputed, the latter referred to the ECJ the following question, pursuant to article 267 of the Treaty on the Functioning of the EU:«*Is Article 1 (2) (e) of the Regulation* [(EC) No 258/97] *of 27 January 1997 to be interpreted as including within its scope foods consisting of whole animals intended to be consumed as such or does it apply only to food ingredients isolated from insects?*».
In the context of the legal proceedings before the ECJ, France and Italy were the only Member States which intervened to contribute with their views to the clarification of the issue of EU law in hand. Both argued in favour of the interpretation whereby whole insects intended for human consumption were covered by the scope of Regulation (EC) No 258/97. In particular, one of the main arguments put forward by the French Government in aid of its views was that public health risks associated with consumption of whole insects were no different from those posed by food ingredients isolated from those animals. In addition to that, the Advocate General Bobek, in his quality of *amicus curiae*, issued his own opinion on the matter on 9 July 2020. By analysing the textual wording of Regulation (EC) No 258/97 as well as Regulation (EU) 2015/2283, the Advocate General concluded that whole insects were *not* covered by the scope of the former regulation. His conclusions are essentially based on a literal interpretation of the text of the regulation as well as on considerations revolving around the intent of the EU legislator at the time of its adoption. Amongst others, Bobek concedes that, back in 1997, the EU legislator may have considered a different approach from that reserved to plants, which were covered by the broader category of '*foods and food ingredients consisting of or isolated from plants*', and not included a similar formulation for animals because (European Court of Justice 2020a):
– «*the available animal-based foods seemed to have had a long history of food use at the time*»; and
– in the specific case of insects, their consumption «*was not really on the menu in Europe at that time*».

(continued)

Table 5.2 (continued)

The ECJ ruling

While the conclusions of the Advocate General are per se not binding on the ECJ, they are generally taken in due account by the latter in the elaboration of its rulings. This is what happened in the case under exam where, in response to the question raised by the referring court, the ECJ eventually ruled that: «*Article 1(2)(e) of Regulation (EC) No 258/97(…) must be interpreted as meaning that foods consisting of whole animals intended to be consumed as such, including whole insects,* do not fall *within the scope of that regulation*». As the Advocate General, the ECJ based its decision essentially on a textual interpretation of that provision, noting in particular that:

– whole insects can be hardly defined as *'food ingredients'* – In the absence of a proper EU definition of 'ingredient', the everyday notion of the latter is that of a product / substance intended to be a component of a larger composite end product and which, in principle, is not to be eaten in and by itself. From this perspective, whole insects would rather qualify for as *'foods'*;

– the provision in question makes explicit reference to food ingredients *'isolated'* from animals, thus implying an extraction process which whole animals obviously do not undergo;

– the text of the regulation contains a different and broader wording for other non-animal products – i.e. *'foods and food ingredients consisting of or isolated from*[…]' plants, fungi, algae and micro-organisms – which may encompass foods which are constituted by a single part (e.g. a whole plant).

Another interesting and crucial point emerging from the ruling is that, in the absence of EU rules regulating whole insects as food pending the application of Regulation (EC) No 258/97, EU Member States had the possibility, under EU law, to fill that legal void by adopting their own national rules, including the prohibition of, or restrictions to, their commercialisation on public health grounds. In so doing, this interpretation seems to provide *ex-post* legal backing to the different approaches taken by EU Member States in this area during the transition from the old to the new novel food regime (see further Sect. 5.1.1.2).

5.1.1.2   From the Old Novel Food Regime to the New One: The Impact on the EU Internal Market

Confronted with the uncertainty of the actual legal status of edible insects under EU law above described, during the transition from the old novel food regime to the new one (indicatively, from 2014 to 2018) several EU Member States gradually started to develop their own interpretation of the matter and established their policy and enforcement approach in accordance with that.

Overall, within the EU three main clusters of countries can be identified on this basis. However, it should be noted that literature review shows that, in certain instances, national approaches have evolved, often over a quite limited span of time, owing to the interaction of different factors and drivers (e.g. science, law, political and/or business pressure etc.). This is the case of Finland, for instance, where the national competent authorities, with a legally dubious change of mind, passed from considering whole insects and their preparations as novel foods to non-novel (Paganizza 2019). Conversely, in Sweden the national competent authorities have permitted the trading of insects for human consumption only after the issuance of the ECJ ruling of October 2020 referred to above.

First Cluster   The first cluster comprises countries in which whole insects intended for human consumption have been considered, for regulatory purposes, to be out of the scope of Regulation (EC) No 258/97 and, consequently, as not novel. Belgium, the Netherlands, Austria, Finland, Sweden and the United Kingdom are amongst the most prominent representatives of this cluster (Lähteenmäki-Uutela and Grmelová 2016; Paganizza 2016; Belluco et al. 2017; Lähteenmäki-Uutela et al. 2018; Lotta 2019; Grabowski et al. 2019: Monteiro et al. 2020).

Belgium, in particular, was the first EU Member State that entered the unchartered territory of policing edible insects. This notably happened following an ad hoc risk assessment performed by the national risk assessor in September 2014, thus before the risk profile on safety of insects as food and feed issued by EFSA (see Sect. 2.5) (Scientific Committee of the Federal Agency for the Safety of the Food Chain 2014). To this effect, during that same year the Belgian authorities published a guidance document to the benefit of the insect business sector illustrating the main elements of the national policy towards the rearing and commercialisation of whole insects and their preparations, with a view to mitigating the regulatory uncertainty surrounding their marketability at EU level. Whilst, originally, the guidance at issue listed ten species which could be reared and commercialised on the national market under specific conditions, its latest version admits only those for which a novel food application was submitted before 1 January 2019, in line with the transitional arrangements foreseen under Regulation (EU) 2015/2283 (see further Sect. 5.1.1.3). These are currently three and notably the house cricket (*A. domesticus*), the migratory locust (*L. migratoria*) and the yellow mealworm (*T. molitor*) (Agence fédérale pour la sécurité de la chaîne alimentaire (AFSCA) 2018; Service Public Féderal 2018; Monteiro et al. 2020).

In the Netherlands, the rearing and the commercialisation of insects for human consumption have been in place already for quite some time and, in general, tolerated by national authorities (Paganizza 2016). However, with growing public interest towards entomophagy, the Dutch authorities also felt appropriate to investigate the potential health risks associated with the consumption of mass-reared insects already in 2014 focussing on the most commercially exploited species at country level (i.e. *L. migratoria*, *T. molitor* and *A. diaperinus*) (Netherlands Food and Consumer Product Safety Authority 2014).

In the United Kingdom, commercial sale of edible insects is also a relatively common practice. For instance, Chinese yellow scorpions – although the latter are arthropods but not strictly insects – have been sold on the national market since the early 1990s coated in chocolate, alcohol or lollies, while roasted crickets and giant ants' eggs are more recent delicacies (Laurenza and Carreño 2015; Finardi and Derrien 2016). In the course of this decade, on a few occasions national authorities have requested businesses of the insect sector to provide data, notably to determine species being on the market and possibly prove their safe use as food before 1997 (Bouckley 2011; Paganizza 2016; Lotta 2019). Based on the research of Grabowski et al. (2019), the United Kingdom has adopted a tolerance policy similar to that of Belgium in order to deal with edible insects marketed in its territory during the application of Regulation (EC) No 258/97, although the national competent authorities did not provide any specific risk assessment, guidance or list of permitted species. With the formal exit of the United Kingdom from the EU, it is however uncertain whether and to what extent this country will opt for maintaining unaltered the EU novel food regime which it has followed up to now.

Other than that, Grabowski et al. (2019) refer that Finland currently permits the commercialisation of six insect species (*A. domesticus*, *G. sigillatus*, *L. migratoria*, *A. diaperinus*, *T. molitor* and *A. mellifera*). All these species, with the exception of *A. mellifera*, are also allowed to be traded in Austria alongside *G. assimilis*, *Gryllus bimaculatus*, *S. americana* and *gregaria*.

This considered, with a view to better understanding how the EU market evolved over the past few years, it is interesting to compare the information on insect species farmed and/or imported for human consumption in the EU which EFSA used to perform its scientific assessment back in 2015 with the more recent research conducted by Grabowski et al. 2019 at Member State level (Table 5.3). This comparison shows that, during the transition period from the previous novel food regime to the one currently in place, few Member States seem to have allowed, under national rules, the production and commercialisation of additional insect species (for instance, *A. mellifera* in Finland and *G. sigillatus* in Austria and again in Finland), thus confirming predictions in that sense formulated by some authors, including Lähteenmäki-Uutela and Grmelová (2016).

Second Cluster Conversely, the second cluster of countries includes EU Member States which have embraced a more restrictive approach towards the possibility of rearing and commercialising insects for human consumption, notably by considering whole insects and their preparations as novel foods requiring prior market authorisation pursuant to Regulation (EC) No 258/97 or, following the repeal of the latter, Regulation (EU) 2015/2283.

Table 5.3 Comparison between insect species farmed or imported for human consumption in the EU in 2015 and insect species allowed by selected Member States in 2019

Insect species	[EU]	[≡]	[▮]	[+]	[≡]
A. domesticus	✓	✓	✓	✓	
G. sigillatus		✓		✓	
G. assimilis		✓			
G. bimaculatus	✓	✓			
G. testaceus	✓				
L. migratora	✓	✓	✓	✓	✓
S. americana	✓	✓			
S. gregaria		✓			
A. diaperinus	✓	✓		✓	✓
T. molitor	✓	✓	✓	✓	✓
Z. atratus	✓	✓			
R. ferrugineus	✓				
B. mori	✓				
A. mellifera				✓	

Source: Elaborated by Montanari, Pinto de Moura and Cunha (2020) based on EFSA (2015) and Grabowski et al. (2019)

Italy, France, Croatia, Czech Republic, Estonia, Ireland, Luxembourg, Poland, Portugal, and Slovenia form part of this group, amongst others (Paganizza 2016; Lotta 2019; Grabowski et al. 2019). However, even within this group, a certain degree of heterogeneity across national policies on edible insects can be found. By way of an example, in Portugal, whilst the rearing and commercialisation of insects has not been allowed during the transition from Regulation (EC) No 258/97 to Regulation (EU) 2015/2283, the national competent authorities have nevertheless permitted, on the national territory, the manufacturing of food products containing insects destined to export with raw materials originating from other EU Member States where their production and commercialisation are lawful (Portugal Insects and Direção-Geral de Alimentação e Veterinária (DGAV) 2019).

In the Czech Republic, retail shops can still sell insect products which have been consumed by consumers of other Member States (IPIFF 2019c). At the other side of the spectrum, the Italian authorities have clearly ruled out also the possibility of any intra-EU trade of insect products intended for the final consumer or for further processing within their own jurisdiction (Ministero della salute 2018). It is in line with this interpretation that, during the EXPO2015 which took place in Milan, the Italian enforcement authorities seized insect products which some Belgian and Dutch manufacturers were showcasing in tasting sessions (Paganizza 2016).

In spite of the above, literature or even a simple research on the web can occasionally provide for evidence that, in some of these countries (e.g. Czech Republic, France, Portugal), there are business operators selling edible insects to consumers through different business models (offline and/or online) (Bednářová et al. 2013; Grabowski et al. 2019; Pippinato et al. 2020). It is also interesting to note that, within this cluster, France was the only country which, through its national food safety agency, performed a proper risk assessment on insects as food (and feed) back in 2015 reaching, overall, the same conclusions as EFSA (Agence nationale de sécurité sanitairede l'alimentation, de l'environnementet du travail (ANSES) 2015).

Third Cluster The third and less numerous cluster comprises EU Member States where, in spite of lack of clear national legislation, the production and the commercialisation of insect species for human consumption is tolerated in certain settings or instances (this is the case of bars and restaurants in Spain and local markets and fairs in Germany) or assessed on a case-by-case basis (e.g. Greece) (Grabowski et al. 2019).

Impact on the EU Internal Market The regulatory heterogeneity observed across the EU during the transition from the old novel food regime to the new one described above makes one wonder about the integrity of the EU internal market. The overall picture resulting from the different national approaches is that of a market which is highly fragmented, subject to rapidly evolving (national) rules and, consequently, characterised by considerable legal uncertainty. This is quite telling if one considers that the very essence of the EU internal market is that businesses, including those pertaining to the agri-food sector, should be able to operate virtually under the same conditions, irrespective of where they are based. Needless to say, in the EU Member States with the more liberal regulatory approaches towards insects as food, the relevant market is steadily developing and/or consolidating at present in terms of number of business operators, range of insect-products and consumer acceptance. As a result, business operators who are located within the jurisdictions of those Member States will be few steps ahead and, therefore, more competitive on the EU market as a whole when the first novel food authorisations regarding insects will be formally adopted at EU level (Lähteenmäki-Uutela and Grmelová 2016) (see further Sect. 5.1.1.4).

From this standpoint, the lack of EU harmonisation in this area – even now that with Regulation (EU) 2015/2283 the legal status of edible insects under EU law has been clarified (see further Sect. 5.1.1.3) – is likely to leave its mark on this emerging sector in the short and medium term by letting that uneven conditions of competition are imposed on its business operators. The economic consequences of this policy approach can be even higher in the case of the insect sector, as opposed to other traditional food sectors, if one considers that the sector is still in its infancy, consists of several small and medium start-ups that are awaiting to enter the EU market and that product portfolios are not highly diversified (see Sect. 3.2.1).

5.1.1.3   Regulation (EU) 2015/2283

Applicable as of 1 January 2018, Regulation (EU) 2015/2283 repealed Regulation (EC) No 258/97 providing the EU with a brand-new legal framework for novel foods. From the standpoint of edible insects, the framework constitutes a significant step forward in terms of legal certainty insofar as it clarifies, in particular, the regulatory status of whole insects, their parts and their preparations as novel foods in the EU (Lähteenmäki-Uutela and Grmelová 2016; Belluco et al. 2017; Rusconi and Romani 2018; Lotta 2019; IPIFF 2019c). This in spite of the fact that the EU inter-institutional negotiations of the new novel food regime addressed other critical food categories in a more structured way (e.g. animal cloning and nanotechnologies) and that the final text of the new regulation contains no single explicit reference to insects as such (Finardi and Derrien 2016). The following paragraphs discuss the novelties of the new regulation notably from the angle of edible insects in terms of material scope, pathways for approval and transitional arrangements.

Scope   Recital 8 of Regulation (EU) 2015/2283 states that «[t]*he scope of this Regulation should, in principle, remain the same as the scope of Regulation (EC) No 258/97. However, on the basis of scientific and technological developments that have occurred since 1997, it is appropriate to review, clarify and update the categories of food which constitute novel foods. Those categories should cover whole insects and their parts* […]».

In accordance with that, while the temporal condition of the definition of novel food did not change as the result of the adoption of the new regulation (see above Sect. 5.1.1.1), article 3 (2) (a) (v) of the latter now lists, amongst the food categories that may qualify as novel, *«food consisting of, isolated from or produced from animals or their parts, except for animals obtained by traditional breeding practices which have been used for food production within the Union before 15 May 1997 and the food from those animals has a history of safe food use within the Union»*.

This wording potentially captures whole insects and their parts under the definition of novel foods pursuant to EU law, besides confirming the most common reading of the regime previously in force whereby food (ingredients) isolated from insects (and their parts) fall also under that definition (Finardi and Derrien 2016; Rusconi and Romani 2018). This also means the insects commercialised for human consumption in the forms above described must undergo a prior safety assessment before being placed on the EU market (Grabowski et al. 2019; Goumperis 2019; Lotta 2019).

Pathways for Approval With regard to the risk assessment that must underpin the placing of insects and products thereof as novel foods on the EU market, it is worth noting that, during the negotiations of the text of Regulation (EU) 2015/2283 and until its entry into application, stakeholders and commentators alike placed high hopes and great emphasis on the new simplified (or 'fast-track') authorisation procedure which the regulation introduced (Paganizza 2016; Lähteenmäki-Uutela and Grmelová 2016; Belluco et al. 2017).

Alongside with the general authorisation procedure (Chapter III, Section I, articles 10–13 of the regulation), the simplified procedure is one of the two possible pathways currently foreseen by the new EU legal framework for the lawful commercialisation of novel foods on the EU market (Chapter III, Section II, articles 14–17).

This procedure essentially consists in a mere 'notification' by an applicant to the European Commission of the intention to place on the EU market certain novel foods – including animals and products thereof and, thus, insects – which, based on the joint reading of article 3 (2) (b) and (c) of the regulation, fulfil the following conditions:

(i) qualify for as 'traditional foods' from a non-EU country, where 'traditional' refers, in particular, to products derived from primary production within the meaning of article 3, point 17, of Regulation (EC) No 178/2002; and

(ii) have a history of safe food use in the country of origin, meaning that *«the safety of the food in question has been confirmed with compositional data and from experience of continued use for at least 25 years in the customary diet of a significant number of people in at least one* [non-EU] *country»*.

From an international trade perspective, the establishment of a simplified authorisation procedure for traditional foods from non-EU countries bears considerable importance as, under the previous novel food regime, only consumption to a significant degree in the EU or any of its Member States before 1997 was deemed relevant to prove safe use of a food (Belluco et al. 2013, 2017) (see also above Sect. 5.1.1.1).

As referred earlier on, the simplified authorisation procedure is intended to be faster than the general authorisation procedure. In principle, the procedure under exam can take up only to a few months (5) from the submission of the application, if no safety objections are raised by EFSA or any Member State, as opposed to an average of 17 months in the case of the general one when risk assessment is needed (which is in fact the rule rather than the exception). Other than that, it only requires the applicant to demonstrate the history of safe consumption in the concerned non-EU country, but not the safety of the food per se through the submission of a full scientific dossier (Rusconi and Romani 2018; Lotta 2019).

For the purposes of the simplified authorisation procedure, an applicant may be a non-EU country or any other interested party, including individual food business operators, trade associations, groups or consortia of businesses based outside or in the EU, in accordance with article 3 (2) (d) of the new regulation. When granted, the approval of a novel food under said procedure is meant to be generic. This constitutes an important change from the regime previously in force and means that the applicant cannot claim any protection of scientific evidence or data used to that effect, nor any related exclusive right to exploit the authorisation commercially, based on article 26 of Regulation (EU) 2015/2283, which is, instead, a possibility foreseen for novel food applications under the general authorisation procedure (see further Sect. 5.1.1.4). Furthermore, under Regulation (EU) 2017/2468, the European Commission adopted technical provisions to ensure its administrative implementation, which were preceded by the publication of a guidance document developed by EFSA to facilitate submissions from potential applicants (respectively, European Commission 2017a and EFSA 2016).

Over the past few years, some authors have identified certain edible insects reared in non-EU countries as suitable candidates for the procedure at issue. For instance, according to Lähteenmäki-Uutela et al. (2018), with regard to crickets in Thailand and mopane worms in South Africa there is currently sufficient information to embark on an EU authorisation. From this perspective, whole insects seem potentially better candidates for the notification procedure in hand as they are more likely to form part of local traditional diets compared to products derived from those animals (e.g. insect proteins) (IPIFF 2019c).

This being said and as it will be shown in more detail under Sect. 5.1.1.4, following the entry into application of Regulation (EU) 2015/2283, none of the novel food applications concerning insect species intended for human consumption which have been submitted to date and are currently pending (n = 12) has been introduced in accordance with the simplified authorisation procedure described above. According to IPIFF (2019b), five novel food applications were presented in relation to insects as traditional foods from non-EU countries though none of them was considered as complete and, consequently, formally validated by the European Commission to be able to move forward with the authorisation process. The same happened to few other applications submitted under the general authorisation procedure as IPIFF (2019b) refers that, by July 2019, the total number of attempted submissions, in fact, amounted to twenty.

In light of the above, one could wonder why insects as traditional foods are currently facing such regulatory hurdles when, as referred earlier on, entomophagy is a common practice in other countries outside the EU, making, in principle, easy to build a case in favour of the suitability of certain insect species as food. In our view, the requirements set by EU legislation for the simplified authorisation procedure may prove particularly challenging in the case of edible insects consumed in and originating from non-EU countries. Effectively, one should consider that demonstrating the history of safe consumption of a traditional food requires the following:

(a) availability of compositional data;
(b) proof of uninterrupted (safe) consumption of that food for at least 25 years as a part of the traditional diet;
(c) proof of the above referred consumption by a significant number of people in the country of origin.

Now it is a fact that, in spite of the diffusion of entomophagy across the globe and the growing scientific interest towards it observed during the last decades, accurate and reliable scientific studies may not be readily available in the majority of the countries and, especially, in the developing part of the world. In fact, amongst the supporting evidence required for the purposes of the simplified authorisation procedure, EFSA recommends to applicants to perform an extensive literature review of studies on human consumption of the traditional food (EFSA 2016).

In addition to that, as most insect species are often consumed by local communities, the question whether the latter can amount to a sufficiently large part of the population in the country of origin of the food for the purpose of the new EU novel food regime arises. Another issue or potential obstacle may derive from having to prove the actual continuity in the consumption of edible insects if one considers that, as several other food products, certain insect species are subject to seasonality or eaten only on special occasions owing to local traditions or their sale price (Belluco et al. 2017; Paganizza 2019; Monteiro et al. 2020).

In addition to the considerations exposed above, there is a further element which may have prevented a greater use of the simplified authorisation procedure in the case of edible insects. This has to do with the scientific uncertainty that surrounds the production and the processing of insects for human consumption in light of the several question marks and areas for further research which EFSA, in its capacity as EU risk assessor, singled out back in 2015. These circumstances, in particular, make quite difficult to believe that a novel food application regarding insects introduced through the simplified authorisation procedure under Regulation (EU) 2015/2283 will trigger no safety objections either from EFSA itself or from any EU Member State. From this perspective, safety concerns around edible insects as novel foods, whatever these may be, seem to be more easily addressable through the submission by concerned applicants of a full dossier in accordance with the general authorisation procedure foreseen by the regulation.

Figures 5.1 and 5.2 illustrate, respectively, the different steps of the simplified and the general authorisation procedures under Regulation (EU) 2015/2283.

Transitional Arrangements   Finally, some of the transitional arrangements set out by Regulation (EU) 2015/2283, which are aimed at ensuring the transition from the

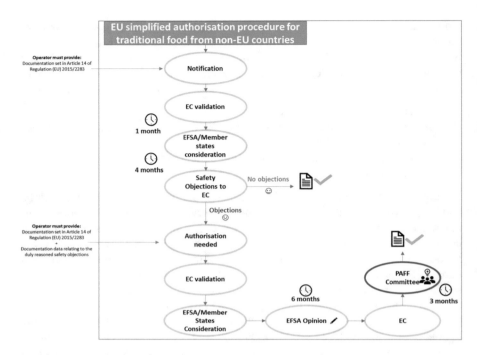

Fig. 5.1 EU simplified authorisation procedure for traditional foods from non-EU countries. (Source: Elaborated by Montanari, Pinto de Moura and Cunha (2020) based on Regulation (EU) 2015/2283)

old novel food regime to the new one, are also relevant for edible insects, even though, also in this case, there is no direct reference to them.

More precisely, pursuant to article 35 (2) of the regulation, foods which *did not fall* under the scope of the previous regime but which are covered by the new framework and which have been lawfully marketed until 1 January 2018 – thus, what else then if not whole insects and their preparations which were marketed on the national markets of certain Member States before the entry into application of Regulation (EU) 2015/2283? – may continue to be commercialised (European Parliament and Council 2015; European Commission 2017a, b):

(i) provided that a novel food application, in accordance with the general or simplified authorisation procedure, has been submitted by 1 January 2019 (as established by Regulations (EU) 2017/2468 and 2017/2469); and
(ii) until a decision in that respect is ultimately taken at EU level.

Therefore, after 1 January 2019, only the insect species and their intended food uses for which a novel food application has been submitted at EU level can be legitimately marketed in the EU.

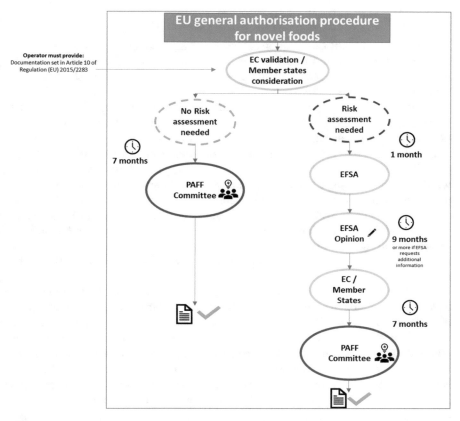

Fig. 5.2 EU general authorisation procedure for novel foods. (Source: Elaborated by Montanari, Pinto de Moura and Cunha (2020) based on Regulation (EU) 2015/2283)

Notwithstanding the above, it would seem that not all Member States (notably, France, Czech Republic, and some Landers in Germany) have been allowing food insect business operators meeting the conditions of article 35 (2) of Regulation (EU) 2015/2283 to fully benefit from the transitional measures under their jurisdictions. While in some cases national practices are based on the argument that whole insects and their preparations *did fall* into the scope of Regulation (EC) No 258/97, in others scientific uncertainty has been invoked and, as result, the precautionary principle set in article 7 of Regulation (EC) No 178/2002 has been applied (IPIFF 2019a, b, c, d, e).

The following section (5.1.1.4) analyses the novel food applications pending by August 2020, taking into account, amongst others, whether their submission has been made in the context of the transitional arrangements provided by new novel food regime.

5.1.1.4   Pending Authorisations

Following the entry into application of the new novel food regime, no insect species has been expressly authorised for human consumption as of yet and, therefore, included in the EU novel food catalogue established by Regulation (EU) 2017/2470 (European Commission 2017c).

There are however at present (August 2020) twelve different applications which are pending at EU level. Table 5.4 provides a detailed overview of the EU applications submitted so far based on the analysis of the information which is publicly available, whilst Fig. 5.3 shows the number of pending applications per insect species. Of those twelve ten are currently undergoing risk assessment at EU level, while for the remaining two EFSA did not consider the scientific dossier to be complete. Presumably, the former are therefore the first applications on which, by the end of 2020 or, most likely, in 2021 an EU authorisation decision may be taken.

Based on the information collected, some general observations can be made as regards actual and potential hindrances in the current authorisation process of edible insects as well as on how the insect market in the EU is likely to evolve in the near future.

First of all, as already referred to above, all outstanding novel food applications have been submitted in accordance with the general authorisation procedure. Taking into account data from IPIFF (2019b) mentioned earlier, no application concerning insects as traditional foods passed the first stage of the authorisation procedure, as opposed to a success rate of 55% in the case of applications lodged in accordance with the generic authorisation procedure. This ultimately casts serious doubts as to whether the simplified procedure applicable to traditional foods originating from non-EU countries is, in fact, a real option for novel food applicants in the case of edible insects.

Secondly, most applications (n = 10) have been lodged under the specific transitional arrangements set out by Regulation (EU) 2015/2283 and described in the previous section, meaning that the concerned insect species had already been present before 1 January 2018 on one or more national markets within the EU. In addition to that, as the applications submitted under the transitional arrangements had to be submitted before 1 January 2019, it can be then concluded that the average timing of 17 months foreseen by the new regulation for the general authorisation procedure has been already exceeded for all applications whose status is currently more advanced, i.e. those currently undergoing EFSA risk assessment. Indeed, by September 2020 the average duration of such procedures was over 23 months with both the risk assessment and the risk management phases to be completed. The administrative phase at the beginning of the procedure has proved particularly lengthy with an average duration of 8.7 months counting from the moment in which the novel food application was submitted to the date when EFSA was formally requested to issue a scientific opinion by the European Commission. Whilst the reasons of such delays in the novel food procedures cannot be always established based on the information publicly available, they may presumably be the result of different circumstances, including the novelty of the procedure and of the food cat-

Table 5.4 Overview of novel food applications concerning insects under Regulation (EU) 2015/2283 currently pending at EU level (September 2020)

Applicant	Country	Procedure, dossier and status	Species	Developmental stage	Physical condition	Intended use(s)	Transitional measures (art. 35 (2))	Identified health risks	Proprietary data (art.26)
⊕PROTIX	▮	Authorisation NF 2018/0804 – EC request to EFSA for a scientific opinion submitted in August 2019. EFSA opinion initially expected by *May 2020*, but currently suspended as additional data were requested to the applicant in March 2020	A. *domesticus*	Adult	Whole (fresh and dried) and ground (dried)	Several food categories including: bread and bakery wares, pasta, dough, spices, condiments, seasonings breakfast cereals, confectionary, alcoholic beverages, soups, salads, snacks, ready-to-eat meals, processed meat, meat substitutes and food for particular nutritional purposes	Yes	Allergenic properties similar to dust mites, crustaceans and molluscs and others depending on the insect diet	Yes

(continued)

Table 5.4 (continued)

Applicant	Country	Species	Developmental stage	Physical condition	Intended use(s)	Transitional measures (art. 35 (2))	Identified health risks	Proprietary data (art.26)
⬤PROTIX	▮▮	*L. migratoria*	Adult	Whole (fresh and dried) and ground (dried)	Several food categories including: bread and bakery wares, pasta, dough, spices, condiments, seasonings breakfast cereals, confectionary, alcoholic beverages, soups, salads, snacks, ready-to-eat meals, processed meat, meat substitutes and food for particular nutritional purposes	Yes	Allergenic properties similar to dust mites, crustaceans and molluscs and others depending on the insect diet	Yes

Procedure, dossier and status: Authorisation NF 2018/0803 – EC request to EFSA for a scientific opinion submitted in August 2019, EFSA opinion initially expected by *May 2020*, but currently suspended as additional data were requested to the applicant in March 2020

⊕PROTIX	II							
	Authorisation NF 2018/0802 – EC request to EFSA for a scientific opinion submitted in August 2019. EFSA opinion initially expected by *May 2020*, but currently suspended as additional data were requested to the applicant in April 2020	*T. molitor*	Larvae		Whole (fresh and dried) and ground (dried)	Several food categories including: bread and bakery wares, pasta, dough, spices, condiments, seasonings breakfast cereals, confectionary, alcoholic beverages, soups, salads, snacks, ready-to-eat meals, processed meat, meat substitutes and food for particular nutritional purposes	Yes	Allergenic properties similar to dust mites, crustaceans and molluscs and others depending on the insect diet	Yes

(continued)

Table 5.4 (continued)

Applicant	Country	Procedure, dossier and status	Species	Developmental stage	Physical condition	Intended use(s)	Transitional measures (art. 35 (2))	Identified health risks	Proprietary data (art.26)
Protifarm	⬛	Authorisation NF 2018/0125 – EC request to EFSA for a scientific opinion submitted in July 2018. EFSA opinion initially expected by *April 2019*, but currently suspended as additional data were requested to the applicant in June 2020	A. *diaperinus*	Larvae	Whole and ground (fresh or dried)	Several food categories including: breakfast cereals and bars, pasta and noodles, bread, bakery wares, chocolate and chocolate products, processed meat, sauces, soups and broths, food supplements and meat substitutes	Yes	Allergic reactions in subjects sensitive to dust mites, crustaceans and molluscs	Yes

▪	Authorisation NF 2018/0128 – Still under consideration by EFSA for risk assessment	A. domesticus	Adult	Whole and ground (fresh and dried)	Several food categories including: pasta, bakery wares, salads, nut and other spreads, snacks, protein-based products.	Yes	Allergenic reactions in subjects sensitive to dust mites, crustaceans and molluscs. Recommendation not to consume more than 54 g of dried insect to comply with daily amount reported as safe for human health by EFSA for chitin (5 g/day)	No

(continued)

Table 5.4 (continued)

Applicant	Country	Procedure, dossier and status	Species	Developmental stage	Physical condition	Intended use(s)	Transitional measures (art. 35 (2))	Identified health risks	Proprietary data (art.26)
	▪	Authorisation NF 2018/0395 – Still under consideration by EFSA for risk assessment	*L. migratoria*	Nymphs or adults	Whole and ground	Several food categories including: protein-based products pasta, flours, bakery wares, salads, nuts and others spreads, snacks, confectionary, sauces, soups and broths	Yes	Allergic reactions in subjects sensitive to dust mites, crustaceans and molluscs Recommendation not to consume more than 95 g of dried insect to comply with daily amount reported as safe for human health by EFSA for chitin (5 g/day)	No

	T. molitor	Larvae	Whole and ground (fresh and dried)	Several food categories including: flours, doughs, pasta, protein-based products, salads, sauces, soups and broths, bakery wares, spreads, snacks and ready-to-eat food products	Yes	Allergic reactions in subjects sensitive to dust mites, crustaceans and molluscs but also in the absence of cross-sensitivity. Recommendation not to consume more than 143 g of dried insect to comply with daily amount reported as safe for human health by EFSA for chitin (5 g/day)	No
	Authorisation NF 2018/0396 - EC request to EFSA for a scientific opinion submitted in December 2019. EFSA opinion initially expected by *September 2020*, but currently suspended as additional data were requested to the applicant in April 2020						

(continued)

Table 5.4 (continued)

Applicant	Country	Procedure, dossier and status	Species	Developmental stage	Physical condition	Intended use(s)	Transitional measures (art. 35 (2))	Identified health risks	Proprietary data (art.26)
	▪	Authorisation NF 2019/1142 - EC request to EFSA for a scientific opinion submitted in June 2020. EFSA opinion expected by *March 2021*	*T. molitor*	Larvae	Ground	Mealworm flour to be used as an ingredient in a range of food categories, including bakery wares, pasta and processed fruit and vegetables	Not indicated	Allergic reactions in subjects sensitive to crustaceans.	Yes

		T. molitor	Larvae	Whole and ground (dried)	Several food categories including: bread and other bakery products, breakfast cereals, bars, confectionery, salad dressings, ready-to-eat food products and pasta	Yes	Allergic reactions in subjects sensitive to crustaceans.	Yes
II	Authorisation NF 2018/0241 - EC request to EFSA for a scientific opinion submitted in July 2018. EFSA opinion initially expected by *March 2019*, but currently suspended as additional data were requested to the applicant in March 2020							

(continued)

Table 5.4 (continued)

Applicant	Country	Procedure, dossier and status	Species	Developmental stage	Physical condition	Intended use(s)	Transitional measures (art. 35 (2))	Identified health risks	Proprietary data (art.26)
microNutris	▊	Authorisation NF 2018/0260 – EC request to EFSA for a scientific opinion submitted in July 2018. EFSA opinion initially expected by *March 2019*, but additional data requested to the applicant in March 2020	*G. sigillatus*	Adult	Whole and ground (dried)	Several food categories including: bread and other bakery products, breakfast cereals, bars, confectionery, salad dressings, ready-to-eat food products and pasta	Yes	Allergic reactions in subjects sensitive to crustaceans.	Yes

		H. illucens	Larvae	Dried ground meal of the whole insect larvae	As an ingredient in bakery wares and potato-, cereal-, flour and starch-based snacks	Not indicated	Allergic reactions in subjects sensitive to dust mites and crustaceans.	No
	Authorisation NF 2018/0765 – EC request EFSA for a scientific opinion submitted in December 2019. EFSA opinion initially expected by *September 2020*, but currently suspended as additional data were requested to the applicant in March 2020.							

(continued)

Table 5.4 (continued)

Applicant	Country	Procedure, dossier and status	Species	Developmental stage	Physical condition	Intended use(s)	Transitional measures (art. 35 (2))	Identified health risks	Proprietary data (art.26)
SML	✝	Authorisation NF 2018/754 – EC request EFSA for a scientific opinion submitted in July 2020. EFSA opinion expected by April 2021	A. mellifera	Male pupae	Whole	Food and food ingredient in other food products	Yes	Allergic reactions in subjects sensitive to dust mites, crustaceans and molluscs	No

Source: Elaborated by Montanari, Pinto de Moura and Cunha (2020) based on the summary of applications made available by the European Commission (https://ec.europa.eu/food/safety/novel_food/authorisations/summary-applications-and-notifications_en) and documents sourced via EFSA Register of Questions (http://registerofquestions.efsa.europa.eu/roqFrontend/login?0)

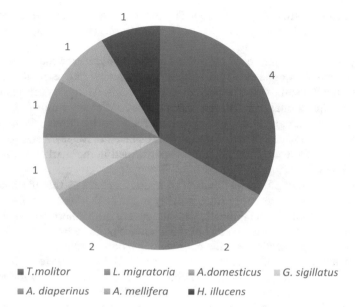

Fig. 5.3 Overview of pending EU novel food applications per insect species (September 2020). (Source: Elaborated by Montanari, Pinto de Moura and Cunha (2020) based on the summary of applications made available by the European Commission (https://ec.europa.eu/food/safety/novel_food/authorisations/summary-applications-and-notifications_en))

egory in hand for the European administration and/or the incompleteness (administrative and/or scientific) of the dossier presented by the applicant. In relation to the scientific evidence submitted with the applications currently subject to risk assessment by EFSA, it is worth noting that in most cases (n = 8) the EU risk assessor requested applicants to provide additional information, which has ultimately resulted in more time for the issuance of the relevant scientific opinions than the standard 9 month-timeframe envisaged by the new regulation (IPIFF 2019c).

Thirdly, in a majority of cases (n = 7) protection of scientific data or evidence supporting the request of authorisation has been invoked. This includes all applications for which EFSA risk assessment is ongoing except the one regarding whole and dried yellow mealworm (*T. molitor*). This element should by no means be underestimated as, in case of a favourable outcome of the authorisation procedure, only the applicant will be able to use and exploit commercially the data and information submitted to support the authorisation request during the timeframe set by Regulation (EU) 2015/2283 for that purpose (5 years). Therefore, even when such authorisations will be granted, one should refrain from jumping to the conclusions that the EU market will be flooded with insects and products thereof destined for human consumption.

Furthermore, considering the specific insect species, the majority of novel food applications (n = 4, i.e. 1/3 of the total) tabled at EU level concerns different food uses of yellow mealworm (*T. molitor*). The latter is followed by the house cricket (*A. domesticus*) and the migratory locust (*L. migratoria*) (with two application dossiers each).

Safety wise, all the proposed novel food applications refer to the possibility that subjects sensitive to crustaceans, molluscs or dust mites may experience allergic reactions after consuming the respective insect species. Allergen properties, alongside with compositional data and relevant legal or food safety limits (e.g. contaminants and microbiological hazards), constitute food specifications which are likely to be reflected in any authorisation decision to be taken at EU level and whose verification will be subject to official controls performed in the market of the various Member States.

Finally, the geographical distribution of applicants reveals that the majority of them are located in Northern Europe, notably in the Netherlands (4), Belgium (3), France (3), Denmark (1) and Finland (1). In the case of the Netherlands and Belgium, in particular, this is certainly a reflex of the tolerance policy adopted by those countries towards the production and the commercialisation of edible insects before the entry into application of Regulation (EU) 2015/2283 (see above Sect. 5.1.1.2). Other than that, the majority of applications (n = 8, i.e. two thirds of the total) has been lodged by individual companies of the insect sector, while in the case of Belgium and Finland applicants are national trade associations which collectively applied for the authorisation of a specific food application of an insect species on behalf of their members.

5.1.2   Regulatory Loopholes

The authorisation decisions of certain insect species and their applications as novel foods which are currently in the pipeline are likely to set out specific requirements in terms of conditions of use (e.g. food categories in which use is allowed and permitted quantities, if relevant), composition (e.g. protein, fats, vitamins, minerals etc.), maximum limits for food-related hazards and consumer information, whose strict observance will be a pre-requisite for their lawful commercialisation on the EU market. Notwithstanding that, once authorised as novel foods, insects will be by default subject to the full set of general EU requirements applying to the production, processing and/or distribution of food. These include, inter alia, the obligations stemming from:

(i) Regulation (EC) No 178/2002 (e.g. food and feed safety, own-controls, traceability, import and export requirements) (European Parliament and Council 2002);

(ii) Regulation (EC) No 852/2004 on the general hygiene of foodstuffs (e.g. compliance with microbiological criteria, establishment of procedures based on HACCP principles, sampling and analysis, registration or approval of the establishments) (European Parliament and Council 2004a);

(iii) Regulation (EU) No 1169/2011 on food information to consumers and Regulation (EC) No 1924/2006 on nutrition and health claims made on foods (European Parliament and Council 2011, 2006).

Against this background, one could wonder whether this general framework is, in fact, per se sufficient – i.e. adequately equipped – to address all the challenges and the needs which the future commercialisation of a brand-new food category across the EU market poses. Looking at the dates of the adoption of the legal acts listed above, it is quite clear that they all precede the emergence of the policy debate surrounding insects as food and, for this reason, do not contain any specific requirements for this purpose. Also, the fact that, over the last few years, the European insect sector and some EU Member States have felt the need to develop detailed guidance covering various aspects of the production chain of insects for human consumption is a clear indicator that the current EU legal framework needs at least to be read through new lenses to be correctly applied to this new food sector (for instance, IPIFF 2019d, 2020c; Grabowski et al. 2019).

Furthermore, in order to establish whether the current EU framework is fit for purpose to regulate the insect sector, consideration should be given not only to safety and quality aspects, but also to the need to ensure comparable market conditions for all business operators of that sector, regardless of whether they are based in or outside the EU, and provide them with an adequate level of legal certainty.

In line with the above, the following sections aim at identifying the main regulatory loopholes which exist at EU level in relation to the production and commercialisation of edible insects whilst discussing their relevance and importance for the proper functioning of the European food insect sector.

5.1.2.1 Food Safety

Protection of human health from foodborne risks is one of the main objectives that led to the establishment of the EU legal framework applicable to food as we know it today. Therefore, if insects are to be eaten as food in future, the EU legislator needs to guarantee that all necessary safety requirements are in place to that effect. From this perspective, there are a few legal acts which will need to be amended in future and whose review is, in our view, rather urgent, taking into account the implications for the health of consumers, on the one hand, and the number of pending novel food authorisations, on the other (see above Sect. 5.1.1.4). In a few areas review of EU legislation aimed at accommodating the needs and specificities of insects as food is already progressing (e.g. hygiene) or even well advanced (e.g. import).

Hygiene and Microbiological Criteria The introduction of specific hygiene requirements for insects as a stand-alone food category of animal origin in Annex III of Regulation (EC) No 853/2004 (European Parliament and Council 2004b) and of specific microbiological criteria within Regulation (EC) No 2073/2005 (European Commission 2005) appear to be particularly important for setting minimum harmonised mandatory requirements for the production and the commercialisation of insects for human consumption on the EU market (Lähteenmäki-Uutela and Grmelová 2016; Lotta 2019).

Indeed, specific hygiene requirements are currently set for a wide range of food of animal origin, including meat, fishery products, live bivalve molluscs and frogs' legs and snails. A similar set of provisions for insects would help operators of that sector to design and implement proper good hygiene manufacturing practices and, in doing so, contribute to a high level of protection of public health in the EU.

This considered, the European Commission did put forward a proposal laying down hygiene rules for insects destined to human consumption late in 2018. However, following extensive consultations with stakeholders at EU level (European Commission 2019a, b, c), eventually the proposal did not find the endorsement of a sufficient number of EU Member States, allegedly because, by largely focussing on regulating substrates for feeding insects, it overlapped with existing EU feed legislation (see further below Sect. 5.3). In the meantime, guidance developed by the competent authorities of certain Member States and by the European insect sector should suffice to this end (see above Sect. 5.1.2) though the adoption of harmonised hygiene rules remains a top priority from a regulatory and market standpoint.

Likewise, as far as microbiological criteria are concerned, EU law currently provides for safety limits for the possible occurrence of specific hazards in different food categories. The latter include crustaceans and molluscan shellfish for the which a hazard such as *Salmonella* must be monitored during processing and its absence guaranteed throughout the shelf-life of the food. The proven biological similarities that exist between insects, on the one hand, and crustaceans and molluscs, on the other, lead to think that safety limits should be consistently set also for the former (IPIFF 2020a). In terms of legislative approach, it is the realistic that, if any, microbiological limits will be set directly in the decisions on novel food authorisations which should be soon adopted. However, at a later stage an amendment to Regulation (EC) No 2073/2005 may be considered by the EU legislator to cover insects as a whole food category.

Contaminants   As discussed under Sect. 2.5, rearing of insects raises concerns about the possible occurrence, during manufacturing and processing, of contamination events with substances which present a risk for public health. This is the case of heavy metals (e.g. cadmium and lead), while other contaminants (e.g. aflatoxins) seem to give rise to less serious concerns. As for microbiological criteria, also here, to the extent which is necessary, maximum limits for specific contaminants are likely to be set by each single novel food authorisation. Yet, in the long run, inclusion of insects under the Annex of Regulation (EC) No 1881/2006, which lists maximum levels for various combinations of contaminants and food categories (European Commission 2006), seems the most logical step.

Maximum Residues of Veterinary Medicines   The prospect of the EU market opening up to the production of edible insects makes think that, inevitably, sooner or later such a production will be scaled up and, thus, resort to agricultural inputs commonly used for rearing food-producing animals, including veterinary medicines. This considered, an update of Regulation (EU) No 37/2010, which sets the maximum residue limits for the occurrence of pharmacologically active substances that can be present in food of animal origin, is soon needed to regulate the specific situation of mass-farming of insects for human consumption (European Commission 2010).

Consumer Information Regarding food information to be provided to consumers through labelling or other suitable means (e.g. online sale channels), considering the amount of evidence collected to date on the potential allergenicity of insects (see above Sect. 2.5), which is confirmed by the novel food applications currently pending at EU level, it can reasonably be expected that product-specific mandatory allergen warnings will be set by the respective authorisations. However, in the long term, insects will most probably need to be included amongst the categories of substances and ingredients which may provoke allergies and intolerances, which are currently listed in Annex II of Regulation (EU) No 1169/2011.

Import Over the last two decades, globalisation has progressively fostered international trade of agri-food products with the EU playing a leading role in terms of imports and exports. As the EU is one of the largest markets for food products worldwide, the prospect that such a market may soon allow the commercialisation of edible insects has inevitably drawn lot of interest by business operators of the insect sector located in non-EU countries. Therefore, establishing the necessary requirements for the import of edible insects into the EU is another important regulatory step which needs to be taken.

In general terms, article 11 of Regulation (EC) No 178/2002 lays down the overarching principle whereby imports destined to the EU market must comply with all the legal requirements applying to food products which are manufactured within the EU or, alternatively, fulfil the conditions which the EU considers as equivalent or which are set in bilateral trade agreements between the EU and the exporting country (European Parliament and Council 2002). Furthermore, in order to be exported to the EU foods of animal origin – a category which, as it has been shown earlier, would legitimately include also insects today – are subject to a two-tier authorisation system, which involves (Kühn and Montanari 2014):

(i) a general export authorisation to be granted to the non-EU country willing to export to the EU, following the provision of guarantees in terms of national legislation in place and effectiveness of official controls as well as the favourable outcome of an audit carried by the European Commission;

(ii) a list of production and/or processing establishments approved for the export to the EU by the national competent authorities of the exporting country and subject to regular controls by the such authorities.

As insects as food are a relatively recent development, no provisions governing the importation of such products into the EU had been previously laid down.

This loophole was recently addressed by the EU legislator through the introduction of a legal basis – article 20 of Regulation (EU) 2019/626 – for the adoption of a list of non-EU countries authorised to export to the EU dead/live insects and their preparations for human consumption (European Commission 2019b). The very first list of non-EU countries to which the authorisation at issue was granted was adopted in November 2019 with Regulation (EU) 2019/1981 and currently includes Canada, Switzerland and South Korea, which offered sufficient guarantees in terms of food safety (European Commission 2019c). Whilst other non-EU countries may be granted the same authorisation in future, the actual list of establishments which will be allowed to export to the EU by the competent authorities of the respective countries is inherently linked with, and will ultimately depend on, the outcome of the various novel food authorisations which are currently pending at EU level.

5.1.2.2   Food Quality

From the perspective of European consumers, food quality is as much as important as food safety and is often a key factor influencing purchasing decisions at retail level. Unlike food safety, food quality is not specifically defined by EU law. Rather, the latter currently regulates quite a broad range of aspects which relate to quality and which can add value to a food product by differentiating it from others pertaining to the same product category. Currently, organic farming and animal welfare are probably the most relevant food quality aspects which are being discussed or considered at EU level in relation to the production and commercialisation of insects for human consumption.

Organic Insect Farming   Organic farming is one of the areas which, in accordance with the recently launched EU Farm-to-Fork Strategy, needs to be further promoted (European Commission 2020). With regard to rearing of insects, legislative negotiations are already taking place at EU level, amongst others, to apply the organic standards set out under Regulation (EU) 2018/848 to the production of insects as food and feed (European Parliament and Council 2018).

The reason why the adoption of requirements for organic insect farming is particularly urgent for the European insect sector is that few other countries which have equivalence agreements in place with the EU (e.g. USA, Canada, Switzerland), by virtue of such agreements, can already export organic-certified insects to the EU. Conversely, this is not true the other way around as in the EU rules for official organic certification of insect farming are currently missing. In this respect, one should also consider that the economic gain expected from organic insects and insect-based products is estimated to amount to a 50% premium price when compared to the corresponding conventional products.

This considered, in the context of the current EU negotiations, the most controversial points concern essentially (IPIFF 2019e):

- the application of certain general animal welfare requirement to insects (e.g. prevention of cannibalism, stocking density, killing techniques) with no adaptation effort to the natural and behavioural characteristics of such animals (however, with regard to animal welfare standards applicable to conventional insect farming see next sub-section); and
- the obligation to source 100% of organic substrates for the feeding of insects, which, allegedly, would pose implementation challenges in terms of market availability of such feed materials and respective prices.

Animal Welfare in Conventional Insect Farming As discussed earlier above (Sect. 2.6), welfare of insects reared for human consumption is a topic which has started drawing some attention recently. In the EU, the integration of animal welfare requirements into the production of edible insects complements and reinforces the positive image which the respective industry sector has been trying to build in terms of sustainability (IPIFF 2019a). The EU has currently a number of legal acts – some general while others specific to certain animal species – in the area of animal welfare. Their aim, in essence, is to guarantee animal protection on the farm, during transport and at the time of killing as a means to improve both animal health of farmed animals and the quality of food they produce (Council 1998, 2005, 2009). However, as they are currently designed, EU animal welfare requirements apply solely to vertebrate animals, which therefore leaves out insects reared for food (or feed) purposes (Lähteenmäki-Uutela and Grmelová 2016).

Against this background, the scientific uncertainty surrounding insects as sentient beings alongside the lack of knowledge and experience in terms of farming practices make difficult, at this point in time, to argue in favour of the need of setting animal welfare rules for such animals at EU level. In this respect, in our view, it is also pertinent to recall that no rules have been established, or even proposed, to date to ensure the welfare of other invertebrates, namely crustaceans and molluscs, species biologically related to insects, which have a long-standing track record as food and cognitive systems which would make them capable to feel pain (EFSA 2005).

Moreover, the establishment of animal welfare rules in the EU generally encounters a certain level of political and business resistance: if mandatory, such requirements are often perceived as adding (unnecessary) production costs that ultimately reflect on the final price paid by consumers. As such, they may entail an undesirable regulatory burden for business operators, all the more in the case of a sector like the one under exam, which is still in its infancy and is largely composed by microcompanies (see Sect. 3.2.1). Likewise, it should not be forgotten that, in the past, the application of certain animal welfare standards solely to establishments based in the EU – for instance, in the case of the EU ban on conventional cages for laying hens under Directive 1999/74/EC (Council 1999; Chippindale 2010) – generated lots of criticism exactly because products from establishments located in non-EU countries could be sold at cheaper price, thus being more competitive on the EU market.

This considered, in spite of animal welfare being one of the key priorities of the EU Farm-to-Fork Strategy, it is very unlikely that ad hoc rules will be adopted for

insects in the foreseeable time at EU level. Investments in further research and promotion of good welfare practices by the insect sector are the most likely scenarios
that lay ahead for the area under exam.

5.2   Feed

As already referred earlier on in Sect. 3.2.2, the feed insect market in the EU is at a
more mature stage in terms of number of companies active in this business segment
and authorised feed applications. Essentially, the current regulatory framework in
the EU allows, under certain conditions and with some limitations, the use of insects
in animal nutrition for different purposes (notably, as pet food and livestock feed)
and in different forms (i.e. as live animals or as insect-derived products, including
insect fat and proteins).

Considering insects as livestock feed, in particular, the applicable regulatory
framework is somehow the result of a complex mix of EU and national rules.

In accordance with that, certain EU Member States allow, under their national
rules, the use of live insects and certain derived products, notably insect fats, to that
effect (IPIFF 2020a).

Other than that, EU law allows for the use of dead whole insects in a limited
number of cases – notably, under national rules, these can be sold as pet food and in
other specialised markets (e.g. zoo and circus animals) – but not to feed farmed
animals (European Parliament and Council 2009a; European Commission 2011,
2013). This is a prohibition which the European insect sector does not consider
justified and which would like to see it changed (IPIFF 2020a).

Also, specific EU restrictions apply in the case of certain insect by-products,
namely insect proteins, which fall under the general category of processed animal
proteins (PAPs). Indeed, the possibility to feed insect proteins to farmed animals is
currently limited on the EU market as a result of a general prohibition to feed ruminants and other animal species with proteins of animal origin, owing to long-
standing safety concerns associated with the Bovine Spongiform Encephalopathy
(BSE) (Paganizza 2016; Lähteenmäki-Uutela et al. 2018; IPIFF 2020a).

For this reason, in accordance with Regulation (EC) No 999/2001, currently,
insect proteins cannot be fed to ruminants (e.g. cows), pigs and poultry (European
Parliament and Council 2001).

Conversely, since 1 July 2017 the use as feed for aquaculture animals of seven
insect species – *H. illucens, M. domestica, T. molitor, A. diaperinus, A. domesticus,
G. sigillatus* and *G. assimilis* – has been allowed, following the adoption of
Regulation (EU) 2017/893, which amended Regulations (EC) No 999/2001 and
(EU) No 142/2011 (European Commission 2017d). The latter constitutes the very
first specific provision set by the EU legislator with regard to insects in the context
of the agri-food chain. From this perspective, it is therefore a landmark step in view
of the consolidation and public recognition of the insect feed and food sector in
Europe. Furthermore, legislative negotiations are ongoing at EU level with a view

to further relaxing the feed ban on PAPs with regard to insect proteins and other proteins from non-ruminant animals, namely to allow their use for the feeding of poultry and swine animals, which is a regulatory development much awaited by the European insect sector (IPIFF 2019b, 2020a). In addition to that, the EU Farm-to-Fork Strategy referred to earlier above singles out insects as one of the potential alternative feed materials to be fostered with a view to reducing the current EU dependency on critical imports destined to animal nutrition (e.g. soya) (European Commission 2020).

5.3   Feed for Insects

In the EU, insects which are reared for food and feed production fall within the category of 'farmed animals', which, in accordance with article 3 (6) (a) of Regulation (EC) No 1069/2009, encompasses *any animal that is kept, fattened or bred by humans and used for the production of food, wool, fur, feathers, hides and skins or any other product obtained from animals or for other farming purposes* (European Parliament and Council 2009a).

   As such, insects are therefore subject to EU rules which regulate the feeding of livestock, including the general principle enshrined in article 4 (1) (a) of Regulation (EC) No 767/2009 whereby animals can be fed only with safe feed (European Parliament and Council 2009b; Lähteenmäki-Uutela and Grmelová 2016). In accordance with that, in terms of substrates, at present insects may only be fed with feed materials of vegetable origin and/or with some materials of animal origin listed in Annex X of Regulation (EU) No 142/2011 (e.g. milk, eggs). Conversely, the use of certain substrates as animal manure, catering waste or former foodstuffs containing fish and meat are prohibited in the EU (European Commission 2011; Lähteenmäki-Uutela et al. 2018). The possibility to enlarge the range of permitted substrates for the feeding of insects intended for animal nutrition to include former foodstuffs and catering waste is one of the key regulatory priorities of European insect sector to further advance this market segment and is likely to be considered at EU level in the next few years (IPIFF 2019b, 2020a).

5.4   Building an EU Regulatory Framework for Insects as Food and Feed: Review Roadmap

Based on the analysis performed, in particular, under Sects. 5.1.2, 5.2 and 5.3, Table 5.5 provides an overview of the EU legal acts which are currently being reviewed or may need reviewing in the near future in order to duly reflect the specificities of the food and feed insect sector. This includes a qualitative assessment as to whether their review is necessary and/or pressing to enable the sector at issue to function properly (notably, in the case of food), consolidate itself (in particular, in the case of feed) and ultimately thrive in the long run.

Table 5.5 Overview of EU legal acts relevant to insects as food and feed and respective review priorities and status

Aspect	Legal Act	Reasons for review	Priority*	Review status
Insects as food				
Hygiene of food of animal origin	Regulation (EU) No 853/2004	Lack of specific hygiene requirements for insects	+++	On hold
Microbiological criteria	Regulation (EC) No 2073/2005	Lack of specific microbiological criteria for insects	++	Likely to be included in the novel food authorisations
Contaminants	Regulation (EC) No 1881/2006	Lack of specific legal limits for insects	++	Likely to be included in the novel food authorisations
Residues of veterinary medicines	Regulation (EU) No 37/2010	Lack of specific maximum limits for residues active substances to treat insects as food-producing animals	++	Not started
Food information	Regulation (EU) No 1169/2011	Insects not included amongst the food categories that may cause allergic reactions in consumers	++	Likely to be included in the novel food authorisations
Import	Regulation (EU) 2019/626	List of non-EU countries authorised to export insect for human consumption to the EU	N/A	Completed but subject to review as appropriate
Organic	Regulation (EU) 2018/848	Lack of rules for organic certification of insects as food	++	Ongoing
Animal welfare	Directive 98/58/EC	Lack of specific rules for farming, transport and killing insects	+	Not started and unlikely to start
	Regulation (EC) No 1/2005			
	Regulation (EC) No 1099/2009			
Insects as feed				
PAPs	Regulation (EC) No 999/2001	Authorising PAPs from insects to be fed to poultry and swine	+++	Ongoing
Feed substrates	Regulation (EU) No 142/2011	Expanding the range of substrates on which insects for animal feed can be reared	++	Likely to start in 2022
Organic	Regulation (EU) 2018/848	Lack of rules for organic certification as feed	++	Ongoing

*+++ = High priority; ++ = Medium priority; + = Low priority

Chapter 6
Conclusions

The analysis performed in the present work, notably in Chap. 5, reveals the existence of few key regulatory constraints to the production and commercialisation of insects as food on the European market. This is true even after the adoption of a new regime for novel foods in the EU which applies to insects.

The first constraint identified is, in fact, the heritage of the legal uncertainty surrounding the permissibility of whole insects and their preparations, in the context of the previous novel food regime, and which has been to some extent perpetuated with the transitional arrangements envisaged by the new EU novel foods framework. In essence, that uncertainty has given rise to different interpretations and national approaches amongst EU Member States, with some tolerating or allowing under certain conditions, the production and the commercialisation of insects as, food and with others prohibiting. This situation has ultimately resulted in the fragmentation of the EU market for this new food product category and has been fostering the development and consolidation of the respective business sector under uneven conditions of competition, both of which are effects that go against the principles governing the economic policies promoted by the EU.

A second constraint identified concerns the implementation of the two authorisation procedures foreseen under the new novel food regime in relation to insects. Effectively, the relevant administrative practice to date shows that only a few attempts have been done to use the simplified procedure – i.e. the one involving the notification of certain novel foods as traditional foods from non-EU countries – in the case of insects and that none of them has eventually advanced further than the actual submission of an application. As far as the general authorisation procedure is concerned, the applications regarding insects which are currently pending at EU level seem to be facing considerable delays as opposed to the standard timeframes envisaged by EU legislation. Whilst the actual reasons (e.g. administrative, scientific etc.) behind the difficulties observed in the management of the authorisation procedures concerning insects need to be further studied, such procedures constitute

© The Author(s), under exclusive license to Springer Nature
Switzerland AG 2021
F. Montanari et al., *Production and Commercialization of Insects as Food and
Feed*, https://doi.org/10.1007/978-3-030-68406-8_6

at present critical bottlenecks for the further uptake of the food insect market in the EU.

As currently designed, the general authorisation procedure for novel foods involves an additional regulatory constraint for the future development of the food insect market in the EU. This relates to the possibility that, in accordance with article 26 of Regulation (EU) 2015/2283, applicants request the protection of certain information – the so-called 'proprietary data' – contained in and submitted in support of their applications, a protection which, if granted, may last up to 5 years from the authorisation date. Whilst this provision aims – reasonably and legitimately - at protecting business investments towards research and development made by applicants, it consequently prevents other business operators from fully exploiting and benefitting from EU authorisations. In the case of insects, the exact impact of this provision on the relevant market segment will be determinable only once the first novel food authorisations are granted. For the time being, it cannot be ignored that for the majority of the novel food applications regarding insects which are currently pending at EU level applicants have formally requested the protection of proprietary data in hand and that, in case such a protection is granted, this may delay the entrance in the EU market of other food insect business operators still for a few years.

Finally, it has been shown that, owing to the fact that the insect sector is a relatively recent novelty in the context of the agri-food chain, currently EU legislation regulates the production and the commercialisation of insects for human consumption to a very limited extent. Therefore, the review of the EU food law *acquis* is necessary to accommodate the needs and specificities of this new category of food-producing animals. From this standpoint and food safety wise, the current lack of hygiene rules for the production of insects as food under Regulation (EC) No 853/2004 is probably the most critical regulatory loophole which needs to be addressed at EU level in the view of the novel food authorisations which soon will be granted to ensure a level playing field between all the operators of this sector.

Conversely, the EU regulatory framework for the production and the commercialisation of insects as feed is at a much more mature stage and, thus, presents, overall, a more limited number of legal constraints. Still, the extension of the range of substrates on which insects reared for feed production may be fed as well as the categories of food-producing animals to which PAPs from insects can be administered are amongst the key priorities of the respective European business sector for ensuring consolidation and further expansion of this market segment.

Notwithstanding the above, the identification of the main regulatory constraints to the production and commercialisation of insects as food and feed in the EU cannot be fully appreciated if analysed on its own. For this purpose, Table 6.1 provides for a comparative analysis, based on a common set of qualitative and quantitative regulatory or policy indicators, between the regulatory frameworks of the EU and other non-EU countries, notably those discussed in Chap. 4 of the present work.

Overall, this shows that the EU appears to be better placed than the majority of the other countries studied as far as regulating insects as food and feed. In particular, together with Switzerland, the EU is the only market where a dedicated regulatory

Table 6.1 Comparative analysis of the legal frameworks for insects as food and feed in the EU and other non-EU countries studied

Indicators		EU	USA	Canada	Australia	Switzerland	Brazil
Legal clarity over the status of insects as food and feed		++	+	++	++	+++	+
Existence of a dedicated regulatory framework for insects as food		Under development	No	No	No	Yes	No
Existence of a dedicated regulatory framework for insects as feed		No	No	No	No	No	No
Current level of harmonisation at country level		+	++	++	++	++	++
Pre-market approval of insects as	Food	Yes (if novel)	Yes (if not GRAS)	Yes (if novel)	Yes (if novel)	Yes (if novel)	Yes (if novel)
	Feed	Yes (generic approval)	Yes (if not GRAS)	Yes (if novel)	No	Yes (generic approval)	Yes but at level of the operator
Level of scientific substantiation required for authorisation		+++	+++	+++	++	+++	++
Number of insect species formally authorised by competent authorities as	Food (n)	0	0	0	3	3	0
	Feed (n)	7	0	1	N/A	7	N/A
Number of insect species currently being assessed for authorisation by competent authorities	Food (n)	7	0	0	0	0	0
	Feed (n)	0	0	0	N/A	0	N/A

Legend: '+' = Low; '++' = Medium; '+++' = High; 'N/A' = not applicable

framework for insects destined for human consumption is being developed. Also, the EU currently has the highest number of insect species authorised for use as animal feed (n = 7) and undergoing approval as food (n = 7). This quite obviously makes it an attractive market for insect operators.

Moreover, regarding insects intended for human consumption, it is apparent that in the EU the level of legal certainty for insect operators has significantly increased with the adoption of a new novel food regime, whilst the granting of the first EU authorisations should ensure that the remaining regulatory differences across EU Member States are overcome. In our view, the level of legal certainty which characterises at present the EU market is higher than that observed in other non-EU countries where:

(i) insects can have multiple legal status and, thus, be subject to different regulatory regimes (e.g. Australia, Canada and USA);

(ii) the operator can determine the actual status of its insect products though taking full legal responsibility for it (e.g. insect proteins considered as GRAS in the USA).

In light of the above, it is expectable that the EU will play a leading role at global level in ensuring that insects are fully integrated into the current agri-food production systems and that, in the long run, they are considered as conventional food or feed. Overall, when compared to the majority of the other Western countries studied here, the current EU policy vis-à-vis insects stands out as more structured, innovation-driven and liberal, though without disregarding food and feed safety. From this perspective, this approach seems to break with the more precautionary and, thus, more restrictive regulatory choices made by the EU legislator not too long ago with regard to other innovative food and feed applications involving genetic engineering and biotechnology.

References

Books, Articles, Research Papers and Studies

Aarts KWP (2020) How to develop insect-based ingredients for feed and food? A company's perspective. J Insects Food Feed 6(1):67–68

Bale JS, van Lenteren JC, Bigler F (2008) Biological control and sustainable food production. Philos Trans R Soc 363:761–776

Batrim J (2017) Insect farming and consumption in Australia – opportunities and barriers. Zoologist 39(1):26–30

Bednářová M, Borkovcová M, Mlček J, Rop O, Zeman L (2013) Edible insects – species suitable for entomophagy under condition of Czech Republic. Acta Universitatits Agriculturae et Silvicolturae Mendelianae Brunensis LXI(3):587–593

Belluco S, Losasso C, Maggioletti M, Alonzi CG, Paoletti MG, Ricci A (2013) Edible insects in a food safety and nutritional review: a critical review. Compr Rev Food Sci Food Saf 12:296–313

Belluco S, Halloran A, Ricci A (2017) New protein sources and food legislation: the case of edible insects and EU law. Food Secur 9:803–814

Bequaert J (1921) Insects as food. How they have augmented the food supply of mankind in early and recent times. J Nat Hist 21:191–200

Bodenheimer FS (1951) Insects as Human Food. W. Junk, The Hague

Bouckley B (2011) 'Edible insects' anyone? Asks FSA, Food Manufacture, 15 Aug 2011. https://www.newfoodmagazine.com/news/18791/fsa-asks-food-companies-for-information-on-edible-insects/

Boyd MC (2017) Cricket soup: a critical examination of the regulation of insects as food. Yale Law Policy Rev 36(2017):17–81

Byrne J (2019) Portuguese feed-insect start-up to build first full scale production unit in 2019, Feednavigator, 5 Feb 2019. Available at https://www.feednavigator.com/Article/2019/02/05/Portuguese-insect-feed-start-up-to-build-first-full-scale-plant#. Last accessed 15 Apr 2020

Caparros MR, Gierts C, Blecker R, Brostaux Y, Haubruge E, Alabi T, Francis F (2016) Consumer acceptance of insect-based alternative meat products in Western countries. Food Qual Prefer 52:237–243

Cardoso S, Letra Mateus T, Lopes J (2020) Porque são os insetos uma nova fonte de alimentos? Parte I. Tecnoalimentar 24:37–39

Chen PP, Wongsiri S, Jamyanya T, Rinderer TE, Vongsamanode S, Matsuka M, Sylvester HA, Oldroyd BP (1998) Honey bees and other insects used as food in Thailand. Am Entomol Spring 44:4–29

Chen X, Feng Y, Chen Z (2009) Common edible insects and their utilization in China. Entomol Res 39:299–303

Cheung LT, Moraes SM (2016) Inovação no setor de alimentos: insetos para consumo humano, Interações. Campo Grande 17(3):503–515

Chippindale N (2010) The 2012 EU ban on conventional cages and its effect. Nuffield Farming Scholarships Trust, A BEMB (R&E) Trust Award

Coutinho J (2017) Insects as a legitimate food ingredient: barriers & strategies. Master thesis, Fundação Getulio Vargas, Escola Brasileira de Administração Pública e de Empresas, Rio de Janeiro

Cunha LM, Ribeiro JC (2019) Sensory and consumer perspectives on edible insects. In: Sogari G et al (eds) Edible insects in the food sector, pp 57–71

Cunha LM, Moura AP, Costa-Lima R (2014) Consumers' associations with insects in the context of food consumption: comparisons from acceptors to disgusted. In: Oral presentation at the conference 'Insects to Feed the World', 14–17 May 2014, The Netherlands

de Conconi JRE (1982) Los insectos como fuente de proteinas en el futuro. Editorial Limuas, México

De Foliart GR (1992a) Insects as human food. Crop Prot 11(5):395–399

De Foliart GR (1992b) Insects: an overlooked food resource. In: Adventures in entomology. Sandhill Crane Press, Gainesville, pp 45–48

Deroy O, Reade B, Spence C (2015) The insectivore's dilemma, and how to take the West out of it. Food Qual Prefer 44:44–55

Dobermann D, Swift JA, Field LM (2017) Opportunities and hurdles of edible insects for food and feed. Nutr Bull 42:293–308

Dreon AL, Paoletti M (2009) The wild food (plants and insects) in Western Friuli local knowledge (Friuli-Venezia Giulia, North Eastern Italy). Contrib Nat Hist 12:461–488

Eisemann CH, Jorgensen WK, Merrit DJ, Rice MJ, Cribb BW, Webb PD, Zalucki MP (1984) Do insects feel pain? A biological view. Experientia 40:164–167

Elorinne A-L, Niva M, Varatinen O, Vaisanen P (2019) Insect consumption among vegans, non-vegan vegetarians, and omnivores. Nutrients 11:292–306

European Parliament (2015) Food safety regulation and policy in the United States. Directorate-General for Internal Policies, Study for the ENVI Committee. Available at https://www.europarl.europa.eu/RegData/etudes/STUD/2015/536324/IPOL_STU(2015)536324_EN.pdf

Evans J, Alemu MH, Flore R, Frøst MB, Halloran A, Jensen AB, Maciel-Vergara G, Meyer-Rochow VB, Münke-Svendsen C, Olsen SB, Payne C, Roos N, Rozin P, Tan HSG, van Huis A, Vantomme P, Eilenberg J (2015) Entompohagy: an evolving terminology in need of review. J Insects Food Feed 1(4):293–305

FAO (2017) The future of food and agriculture – trends and challenges. Food and Agriculture Organization of the United Nations, Rome

Finardi C, Derrien C (2016) Novel food: where are insects (and feed) in regulation 2015/2283? Eur Food Feed Law Rev 2:119–129

Finkel MD, Rojo S, Roos N, van Huis A, Yen LA (2015) The European Food Safety Authority scientific opinion on a risk profile related to production and consumption of insects as food and feed. J Insects Food Feed 1(4):245–247

Foliart D (1999) Insects as food: why the Western attitude is important. Annu Rev Entomol 44:21–50

Gasco L, Biasato I, Dabbou S Schiavone A, Gai F (2019) Quality and consumer acceptance of products from insect-fed animals. In: Sogari G et al (eds) Edible insects in the food sector, pp 73–86

Gaulkin S (2020) Calls for regulatory approval of edible insects, the regulatory review, 6 May 2020. Available at https://www.theregreview.org/2020/05/06/gaulkin-calls-regulatory-approval-edible-insects/. Last accessed 27 June 2020

Goumperis T (2019) Insects as food: risk assessment and their future perspective in Europe. In: Sogari G et al (eds) Edible insects in the food sector, pp 1–10

Grabowski NT, Klein G (2016) Microbiology of processed edible insect products – results of a preliminary survey. Int J Food Microbiol 243:103–107

Grabowski NT, Klein G, Martinez AL (2013) European and German food legislation facing uncommon foodstuffs. Crit Rev Food Sci Nutr 53:787–800

Grabowski NT, Ahlfeld B, Lis KA, Jansen W, Kehrenberg C (2019) The current legal status of edible insects in Europe. Berliner and Münchener Tierärztliche Wochenschifrt 132(Heft 5/6):295–311

Gracer D (2010) Filling the plates: serving insects to the public in the United States. Forest insects as food: a global review. In: Durst PB et al (eds) Forest insects as food: humans bite back. FAO, Bangkok, pp 217–220

Grmelová N, Sedmidusbsky T (2017) Legal and environmental aspects of authorizing edible insects in the European Union. Agric Econ Czech 63(9):393–399

Halloran A, Münke C (2014) Discussion paper: regulatory frameworks influencing insects as food and feed. FAO. Available at http://www.fao.org/edible-insects/39620-04ee142dbb758d9a521c619f31e28b004.pdf

Halloran A, Vantomme P, Hanboonsong Y, Ekesi S (2015) Regulating edible insects: the challenge of addressing food security, nature conservation, and the erosion of traditional food culture. Food Secur 7:739–746

Hartmann C, Siegrist M (2016) Becoming and insectivore: results of an experiment. Food Qual Prefer 51:118–122

House J (2017) Consumer acceptance of insect-based foods in the Netherlands: academic and commercial implications. Appetite 106:47–58

Illgner P, Nel E (2000) The geography of edible insects in sub-Saharan Africa; a study of Mopane caterpillar. Geogr J 166:336–351

Intergovernmental Science-Policy Platform on Biodiversity and Ecosystem Services (IPBES) (2016) Assessment report on pollinators, pollination and food production. In: Potts SG, Imperatriz-Fonseca VL, Ngo HT (eds) Secretariat of the Intergovernmental Science-Policy Platform on Biodiversity and Ecosystem Services, Bonn. Available at https://ipbes.net/assessment-reports/pollinators. Accessed 29 June 2020

Johnson DV (2010) The contribution of edible forest insects to human nutrition and to forest management. Forest insects as food: a global review. In: Durst PB et al (eds) Forest insects as food: humans bite back. FAO, Bangkok, pp 5–22

Jongema Y (2017) List of edible insects of the world. Available at https://www.wur.nl/en/Expertise-Services/Chair-groups/Plant-Sciences/Laboratory-of-Entomology/Edible-insects/Worldwide-species-list.htm. Accessed 29 June 2020

Kampichler C, Bruckner A (2009) The role of microarthropods in terrestrial decomposition: a meta-analysis of 40 years of litterbag studies. Biol Rev 84:375–389

Katayama N, Ishikawa Y, Takaoki M, Yamashita M, Nakayama S, Kiguchi K, Kok R, Wada H, Mitsuhashi J (2007) Entomophagy: a key space agriculture. Adv Space Res 2007:1–5

Kirova M, Montanari F, Ferreira I, Pesce M, Albuquerque JD, Montfort C, Neirynck R, Moroni J, Traon D, Perrin M, Echarri J, Arcos PA, Lopez ME, Pelayo E (2019) Research for AGRI committee – megatrends in the agri-food sector. European Parliament, Policy Department for Structural and Cohesion Policies, Brussels

Knutsson S, Munthe C (2017) A virtue of precaution regarding the moral status of animals with uncertain sentience. J Agric Environ Ethics 30:213–224

Koeleman E (2017) New proteins – Canadian approval for insects in salmon feed, 27 February 2017. Available at https://www.allaboutfeed.net/New-Proteins/Articles/2017/2/Canadian-approval-for-insects-in-salmon-feed-99434E/. Last accessed 15 Apr 2020

Kostecka J, Konieczna K, Cunha LM (2017) Evaluation of insect-based food acceptance by representatives of Polish consumers in the context of natural resources processing retardation. J Ecol Eng 18(2):166–174

Kühn C, Montanari F (2014) Importing food into the EU. In: van der Meulen B (ed) EU food law handbook, pp 443–470

Kuljanic N, Gregory-Manning S (2020) What if insects were on the menu in Europe? Scientific Foresight Unit (STOA), European Parliament Research Service. Available at https://www. europarl.europa.eu/RegData/etudes/ATAG/2020/641551/EPRS_ATA(2020)641551_EN.pdf

La Barbera F, Verneau F, Amato M, Grunert K (2018) Understanding Westerners' disgust for the eating of insects: the role of food neophobia and implicit associations. Food Qual Prefer 64:120–125

Lähteenmäki-Uutela A, Grmelová N (2016) European law on insects in food and feed. EU Food Feed Law Rev 1:2–8

Lähteenmäki-Uutela A, Grmelová N, Hénault-Ethier L, Deschamps MH, Vandenberg GW, Zhao A, Zhang Y, Yang B, Nemane V (2017) Insects as food and feed: laws of the European Union, United States, Canada, Mexico, Australia, and China. EU Feed Food Law Rev 1:22–36

Lähteenmäki-Uutela A, Hénault-Ethier L, Marimuthu SB, Talibov S, Allen RN, Nemane V, Vandenberg GW, Józefiak D (2018) The impact of the insect regulatory system on the insect marketing system. J Insect Food Feed 4(3):187–198

Lamsal B, Wang H, Pinsirodom P,·Dossey AT (2019) Applications of insect-derived protein ingredients in food and feed industry. J Am Oil Chem Soc 96:105–123

Laurenza E, Carreño I (2015) Edible insects and insect-based products in the EU: safety assessments, legal loopholes and business opportunities. Eu J Risk Regul 2(2015):288–292

Le Beau A (2015) Insect protein: what are the food safety and regulatory challenges? 1 August 2015. Available at https://burdockgroup.com/insect-protein-what-are-the-food-safety-and-regulatory-challenges/#_ftn19. Last accessed 27 June 2020

Lesnik JJ (2017) Not just a fallback food: global patterns of insect consumption related to geography, not agriculture. Am J Hum Biol 29:e22976

Loo S, Sellbach U (2013) Eating (with) insects: insect gastronomies and upside-down ethics. Parallax 19:12–28

López S, González M, Goldarazena A (2011) Spread of the yellow-legged hornet Vespa velutina nigrithorax du Buysson (Hymenoptera: Vespidae) across Northern Spain. OEPP/EPPO Bull 41:439–441

Lotta F (2019) Insects as food: the legal framework. In: Sogari G et al (eds) Edible insects in the food sector, pp 105–118

Mancini A, Moruzzo R, Riccioli F, Paci G (2019) European consumers' readiness to adopt insects as food. A review. Food Res Int 122:661–678

Manunza L (2019) Casu Marzu: a gastronomic genealogy. In: Halloran et al (eds) Edible insects in sustainable food systems, pp 139–145

Medeiros CE (2013) Insects as human food: an overview. Amazôn, Revista de Antropologia 5(3) Especial:562–582

Meticulous Research (2019) Edible insects market – global opportunity analysis and industry forecast (2019–2030). Available at https://www.meticulousresearch.com/product/edible-insects-market-forecast/

Monteiro S, Carrilho E, Oliveira C (2020) Insetos – Alimentos para o futuro. Riscos e Alimentos 19:17–22

Neves ATSG (2015) Determinants of consumers' acceptance of insects as food and feed: a cross-cultural study. MSc dissertation, Faculty of Sciences, University of Porto

Nonaka K (2009) Feasting on insects. Entomol Res 39:304–312

Noriega JA, Hortal J, Azcárate FM, Berg MP, Bonada N, Briones MJI, Del Toro I, Goulson D, Ibanez S, Landis DA, Moretti M, Potts SG, Slade EM, Stout JC, Ulyshen MD, Wackers FL, Woodcock BA, Santos AMC (2018) Research trends in ecosystem services provided by insects. Basic and Appl Ecol 26:8–23

Oonincx DGAB, de Boer IJM (2012) Environmental impact of the production of mealworms as a protein source for humans – a life cycle assessment. PLoS One 7(12):e51145. pp 1–5

Paganizza V (2016) Eating insects: crunching legal clues on entomophagy. Rivista di diritto alimentare, Anno X, numero 1, Gennaio- Marzo, pp 16–41

Paganizza V (2019) Bugs in law. Wolters Kluwer, Milano

Payne C, Caparros MR, Dobermann D, Frédéric F, Shockley M, Sogari G (2019) Insects as food in the global north – the evolution of the entomophagy movement. In: Sogari G et al (eds) Edible insects in the food sector, pp 11–26

Persistent Market Research (2019) Global market study on edible insects for animal feed: increasing commercialization witnessed of insect-based protein for Aquafeed. Available at https://www.persistencemarketresearch.com/market-research/edible-insects-for-animal-feed-market.asp

Pippinato L, Gasco L, Di Vita G, Mancuso T (2020) Current scenario in the European edible-insect industry: a preliminary study. J Insects Food Feed (in press)

Poma G, Cuykx M, Amato E, Calaprice C, Focant JF, Covici A (2017) Evaluation of hazardous chemicals in edible insects and insect-based food intended for human consumption. Food Chem 100:70–79

Popoff M, MacLeod M, Leschen W (2017) Attitudes towards the use of insect-derived materials in Scottish salmon feeds. J Insect Food Feed 3:131–138

Rader R et al (2016) Non-bee insects are important contributors to global crop pollination. Proc Natl Acad Sci U S A 11

Raheem D, Carrascosa C, Oluwole OB, Nieuwland M, Saraiva A, Millán R, Raposo A (2018) Traditional consumption of and rearing of edible insects in Africa, Asia and Europe. Critic Rev Food Sci Nutr Taylor and Francis 59:1–20

Reverberi M (2017) Exploring the legal status of edible insects around the world, Food Navigator Asia, 1 February 2017. Available at https://www.foodnavigator-asia.com/Article/2017/02/01/Exploring-the-legal-status-of-edible-insects-around-the-world#

Ribeiro JC, Cunha LM, Sousa-Pinto B, Fonseca J (2018) Allergic risks of consuming edible insects: a systematic review. Mol Nutr Food Res 62(1):170030. pp 1–12

Ribeiro JC, Cunha LM, Sousa-Pinto B, Fonseca J (2019) Potential allergenic risks of entomophagy. In: Sogari G et al (eds) Edible insects in the food sector, pp 87–104

Rogers EM (2003) Diffusion of innovations, 5th edn. Free Press, New York

Roos N, van Huis A (2017) Consuming insects: are there health benefits? J Insect Feed Food 3(4):225–229

Ruby MB, Rozin P (2019) Disgust, sushi consumption, and other predictors of acceptance of insects as food by Americans and Indians. Food Qual Prefer 74:155–162

Rumpold BA, Schlüter (2013) Potential and challenges of insects as innovative source for food and feed production. Innov Food Sci Emerg Technol 17:1–11

Rusconi G, Romani L (2018) Insects for dinner – the next staple food. Eur Food Feed Law Rev 4:335–339

Schabel HG (2010) Forest insects as food: a global review. In: Durst PB et al (eds) Forest insects as food: humans bite back. FAO, Bangkok, pp 37–64

Schelomi M (2015) Why we still don't eat insects: assessing entomophagy promotion through a diffusion of innovations framework. Trends Food Sci Technol 45:311–318

Schlups Y, Brunner T (2018) Prospects for insects as food in Switzerland: a tobit regression. Food Qual Prefer 68(2018):37–46

Schrader J, Oonixcx DGAB, Ferreira MP (2016) North American entomophagy. J Insect Food Feed 2(2):111–120

Sutton MQ (1995) Archaeological aspects of insect use. J Archaeol Method Theory 2:253–298

Tan HSG, Fischer ARH, Tinchan P, Stieger M, Steenbekkers LPA, van Trijp HCM (2016) Insects as food: exploring cultural exposure and individual experience as determinants of acceptance. Food Qual Prefer 52:222–231

van der Merwe NJ, Thackeray JC, Lee-Thorp JA, Luyt J (2003) The carbon isotope ecology and diet of Australopithecus Africanus at Sterkfontein, South Africa. J Hum Evol 44:581–597

van Huis A (2010) Opinion: Bugs can solve food crisis. Scientist Magaz Life Sci

van Huis A (2013) Potential of insects as food and feed in assuring food security. Annu Rev Entomol 58:563–583

van Huis A (2015) Edible insects contributing to food security. Agric Food Secur 2015(4):1–9 van Huis A (2017) Edible insects: marketing the impossible? J Insect Food Feed 3(2):67–68

van Huis A (2020a) Insects as food and feed, a new emerging agricultural sector: a review. J Insect Food Feed 6(1):27–44

van Huis A (2020b) Prospects of insects as food and feed. Org Agric 2020

van Huis A (2020c) Welfare of farmed insects. J Insect Food Feed 5(3):159–162

van Huis A, Oonincx DGAB (2017) The environmental sustainability of insects as food and feed. A review. Agron Sustain Dev 37:43

van Huis A, van Itterbeek J, Klunder H, Mertens E, Halloran A, Muir G, Vantomme P (2013) Edible insects: future prospects for food and feed security. Food and Agriculture Organization of the United Nation, Rome

van Lenteren JC (2006) Ecosystem services to biological control of pests: why are they ignored? Proc Neth Entomol Soc Meet 17:103–111

Van Thielen L, Vermuyten S, Storms S, Rumpold B, Van Campenhout L (2019) Consumer acceptance of foods containing edible insects in Belgium tow year after their introduction to the market. J Insect Food Feed 5:35–44

Varelas V (2019) Food wastes as a potential new source for edible insect mass production for food and feed: a review. Fermentation 5(3):81–100

Verbeke W (2015) Profiling consumers who are ready to adopt insects as a meat substitute in a Western society. Food Qual Prefer 39:147–155

Verbeke W, Spranghers T, De Clerq P (2016) Insects in animal feed: acceptance and its determinants among farmers, agriculture sector stakeholders and citizens. Anim Feed Sci 204:72–87

Watson E (2016) Edible insects – beyond the novelty factor, Food Navigator USA, 27 September 2016. https://www.foodnavigator-usa.com/Article/2016/09/28/Edible-insects-Beyond-the-novelty-factor?utm_source=copyright&utm_medium=OnSite&utm_campaign=copyright. Last accessed 27 June 2020

Wilkinson K, Muhlhausler B, Motley C, Crump A, Bray H, Anken R (2018) Australian consumers' awareness and acceptance of insects as food. Insects 9(44):1–11

Yen LA (2009) Edible insects: traditional knowledge or western phobia? Entomologic Res 39:289–298

Yen LA (2010) Edible insects and other invertebrates in Australia: future prospects. Conference proceeding Food and Agriculture Organisation of the United Nations. www.cabdirect.org

Policy, Legislation and Case-Law

Codex Alimentarius

FAO/WHO Coordinating Committee for Asia (2010) Development of regional standard for Edible Crickets and their products. In: Seventeenth Session Bali, Indonesia, 22–26 November 2010. Available at http://www.fao.org/tempref/codex/Meetings/CCASIA/ccasia17/CRDs/AS17_CRD08x.pdf. Last accessed 27 June 2020

Australia

Food Standards Australia and New Zealand (2020) Record of views formed in response to inquiries – Updated May 2020. Available at https://www.foodstandards.gov.au/industry/novel/novelrecs/Pages/default.aspx. Last accessed 22 June 2020

Belgium

Agence fédérale pour la sécurité de la chaîne alimentaire (2018) Circulaire relative à l'élevage et à la commercialisation d'insectes et de denrées à base d'insectes pour la consommation humaine, 5 November 2018. Available at http://www.afsca.be/professionnels/denreesalimentaires/circulaires/_documents/2018-11-05_omzendbriefinsectenv3FR_clean.pdf. Last accessed 26 July 2020

Service Public Féderal (2018) Etat des lieux concernant la commercialisation des insectes ou des produits à base d'insectes en vue de la consommation humaine sur le marché belge après le 1/1/2018. Available at https://www.health.belgium.be/sites/default/files/uploads/fields/fpshealth_theme_file/2018_01_revision_insects_stateoftheplay_fr.pdf. Last accessed 26 July 2020

Scientific Committed of the Federal Agency for the Safety of the Food Chain and of the Health Council (2014) Food safety aspects of insects intended for human consumption (SciCom dossier 2014/04; SHC dossier n° 9160). Available at http://www.afsca.be/scientificcommittee/opinions/2014/_documents/Advice14-2014_ENG_DOSSIER2014-04.pdf. Last accessed 26 July 2020

Brazil

Anvisa (1999a) Resolução n.° 16, de 30 de abril de 1999 - Aprova o Regulamento Técnico de Procedimentos para registro de Alimentos e ou Novos Ingredientes, constante do anexo desta Portaria, Diário Oficial da União de 3 de dezembro 1999

Anvisa (1999b) Resolução n° 17, de 30 de abril de 1999 – Regulamento técnico que estabelece as diretrizes básicas para avaliação de risco e segurança dos alimentos, Diário Oficial da União de 3 de maio de 1999

Anvisa (2014) Resolução da diretoria colegiada n.° 14, de 28 de março de 2014 – Dispõe sobre matérias estranhas macroscópicas e microscópicas em alimentos e bebidas, seus limites de tolerância e dá outras providências, Diário Oficial da União de 31 de março 2014

Canada

Canada Food Inspection Agency (2019) Comment on: registration requirements for insect-derived livestock feed ingredients. Available at https://www.inspection.gc.ca/about-cfia/transparency/consultations-and-engagement/insect-derived-livestock-feed-ingredients/eng/1557839964825/1557839965086. Last accessed 27 June 2020

Government of Canada (1983) Feeds Regulations (SOR/83-593)

Government of Canada (1985a) Feeds Act, R.S.C., c. F-9

Government of Canada (1985b) Food and Drugs Act, R.S.C., c. F-27

Government of Canada (2019) Food and Drug Regulations C.R.C., c. 870, as last amended

Government of Canada (2009) Annual Report on the Operation of the Canadian Multiculturalism Act, 2007–2008

European Union

Council (1998) Directive 98/58/EC of 20 July 1998 concerning the protection of animals kept for farming purposes. OJ L 221, 8.8.1998, pp 23–27

Council (1999) Directive 1999/74/EC of 19 July 1999 laying down minimum standards for the protection of laying hens. OJ L 203, 3.8.1999, pp 53–57

Council (2005) Regulation (EC) No 1/2005 of 22 December 2004 on the protection of animals during transport and related operations and amending Directives 64/432/EEC and 93/119/EC and Regulation (EC) No 1255/97. OJ L 3, 5.1.2005, pp 1–4

Council (2007) Regulation (EC) No 834/2007 of 28 June 2007 on organic production and labelling of organic products and repealing Regulation (EEC) No 2092/91. OJ L 189, 20.7.2007, pp 1–23

Council (2009) Regulation (EC) No 1099/2009 of 24 September 2009 on the protection of animals at the time of killing. OJ L 303, 18.11.2009, pp 1–30

European Commission (2005) Regulation (EC) No 2073/2005 of 15 November 2005 on microbiological criteria for foodstuffs. OJ L 338 22.12.2005, pp 1–26

European Commission (2006) Regulation (EC) No 1881/2006 of 19 December 2006 setting maximum levels for certain contaminants in foodstuffs. OJ L 364, 20.12.2006, pp 5–24

European Commission (2010) Regulation (EU) No 37/2010 of 22 December 2009 on pharmacologically active substances and their classification regarding maximum residue limits in foodstuffs of animal origin. OJ L 15, 20.1.2010, pp 1–72

European Commission (2011) Regulation (EU) No 142/2011 of 25 February 2011 implementing Regulation (EC) No 1069/2009 of the European Parliament and of the Council laying down health rules as regards animal by-products and derived products not intended for human consumption and implementing Council Directive 97/78/EC as regards certain samples and items exempt from veterinary checks at the border under that Directive. OJ L 54, 26.2.2011, pp 1–254

European Commission (2013) Regulation (EU) No 68/2013 of 16 January 2013 on the Catalogue of feed materials. OJ L 029 30.1.2013, pp 1–64

European Commission (2017a) Implementing Regulation (EU) 2017/2468 of 20 December 2017 laying down administrative and scientific requirements concerning traditional foods from third countries in accordance with Regulation (EU) 2015/2283 of the European Parliament and of the Council on novel foods. OJ L 351, 30.12.2017, pp 55–63

European Commission (2017b) Implementing Regulation (EU) 2017/2469 of 20 December 2017 laying down administrative and scientific requirements for applications referred to in Article 10 of Regulation (EU) 2015/2283 of the European Parliament and of the Council on novel foods. OJ L 351, 30.12.2017, pp 64–71

European Commission (2017c) Implementing Regulation (EU) 2017/2470 of 20 December 2017 establishing the Union list of novel foods in accordance with Regulation (EU) 2015/2283. OJ L 351, 30.12.2017, pp 72–201

European Commission (2017d) Regulation (EU) 2017/893 of 24 May 2017 amending Annexes I and IV to Regulation (EC) No 999/2001 of the European Parliament and of the Council and Annexes X, XIV and XV to Commission Regulation (EU) No 142/2011 as regards the provisions on processed animal protein. OJ L 138, 25.5.2017, pp 92–116

European Commission (2018) The rapid alert system for food and feed – 2018 annual report. Available at https://ec.europa.eu/food/sites/food/files/safety/docs/rasff_annual_report_2018.pdf

European Commission (2019a) Commission REGULATION (EU) .../...of XXX amending Annex III to Regulation (EC) No 853/2004 of the European Parliament and of the Council as regards specific hygiene requirements for insects intended for human consumption, Ref. Ares(2019)382900 – 23/01/2019

European Commission (2019b) Implementing Regulation (EU) 2019/626 of 5 March 2019 concerning lists of third countries or regions thereof authorised for the entry into the European Union of certain animals and goods intended for human consumption, amending Implementing Regulation (EU) 2016/759 as regards these lists. OJ L 131, 17.5.2019, pp 31–50

European Commission (2019c) Implementing Regulation (EU) 2019/1981 of 28 November 2019 amending Implementing Regulation (EU) 2019/626 as regards lists of third countries and regions thereof authorised for the entry into the European Union of snails, gelatine and collagen, and insects intended for human consumption. OJ L 308, 29.11.2019, pp 72–76

European Commission (2020) Communication to the European Parliament, the Council, the European Economic and Social Committee and the Committee of the Regions – A Farm to Fork Strategy for a fair, healthy and environmentally-friendly food system, COM (2020) 381 final, 20 May 2020

European Court of Justice (2005) Judgment of the Court (First Chamber) of 9 June 2005. HLH Warenvertriebs GmbH (C-211/03) and Orthica BV (C-299/03 and C-316/03 to C-318/03) v Bundesrepublik Deutschland, ECLI identifier: ECLI:EU:C:2005:370

European Court of Justice (2019) Entoma SAS v. Ministre de l'Économie et des Finances, Ministre de l'Agriculture et de l'Alimentation, Case C- 526/19, Request of a preliminary ruling (working document), 9 July 2019. OJ C 328, 30.9.2019, p 24

European Court of Justice (2020a) Entoma SAS v. Ministre de l'Économie et des Finances, Ministre de l'Agriculture et de l'Alimentation, Case C- 526/19, Opinion of the Advocate General Bobek, 9 July 2020, ECLI identifier: ECLI:EU:C:2020:552

European Court of Justice (2020b) Entoma SAS v. Ministre de l'Économie et des Finances, Ministre de l'Agriculture et de l'Alimentation, Case C- 526/19, Judgment of the Court (Third Chamber), 1October 2020ECLI identifier: ECLI:EU:C:2020:769

European Food Safety Authority (2005) Opinion on the "Aspects of the biology and welfare of animals used for experimental and other scientific purposes", Scientific Panel on Animal Health and Welfare. EFSA J 292:1–46

European Food Safety Authority (2015) Scientific Opinion on a risk profile related to production and consumption of insects as food and feed, EFSA Scientific Committee. EFSA J 13(10):4257, 60 pp. https://doi.org/10.2903/j.efsa.2015.4257

European Food Safety Authority (2016) Guidance on the preparation and presentation of the notification and application for authorisation of traditional foods from third countries in the context of Regulation (EU) 2015/2283. EFSA J 14(11):4590, 16 pp https://doi.org/10.2903/j.efsa.2016.4590

European Parliament and Council (1997) Regulation (EC) No 258/97 of 27 January 1997 concerning novel foods and novel food ingredients. OJ L 43, 14.2.1997, pp 1–6

European Parliament and of the Council (2001) Regulation (EC) No 999/2001 of 22 May 2001 laying down rules for the prevention, control and eradication of certain transmissible spongiform encephalopathies. OJ L 147, 31.5.2001, pp 1–40

European Parliament and Council (2002) Regulation (EC) No 178/2002 of 28 January 2002 laying down the general principles and requirements of food law, establishing the European Food Safety Authority and laying down procedures in matters of food safety. OJ L 31, 1.2.2002, pp 1–24

European Parliament and Council (2004a) Regulation (EC) No 852/2004 of 29 April 2004 on the hygiene of foodstuffs. OJ L 139, 30.4.2004, pp 1–54

European Parliament and Council (2004b) Regulation (EC) No 853/2004 of 29 April 2004 laying down specific hygiene rules for food of animal origin. OJ L 139, 30.4.2004, pp 55–205

European Parliament and Council (2006) Regulation (EC) No 1924/2006 of 20 December 2006 on nutrition and health claims made on foods. OJ L 404, 30.12.2006, pp 9–25

European Parliament and Council (2009a) Regulation (EC) No 1069/2009 of 21 October 2009 laying down health rules as regards animal by-products and derived products not intended for human consumption and repealing Regulation (EC) No 1774/2002. OJ L 300, 14.11.2009, pp 1–33

European Parliament and Council (2009b) Regulation (EC) No 767/2009 of 13 July 2009 on the placing on the market and use of feed, amending European Parliament and Council Regulation (EC) No 1831/2003 and repealing Council Directive 79/373/EEC, Commission Directive 80/511/EEC, Council Directives 82/471/EEC, 83/228/EEC, 93/74/EEC, 93/113/EC and 96/25/EC and Commission Decision 2004/217/EC. OJ L 229, 1.9.2009, pp 1–28

European Parliament and Council (2011) Regulation (EU) No 1169/2011 of 25 October 2011 on the provision of food information to consumers, amending Regulations (EC) No 1924/2006 and (EC) No 1925/2006 of the European Parliament and of the Council, and repealing Commission Directive 87/250/EEC, Council Directive 90/496/EEC, Commission Directive 1999/10/EC, Directive 2000/13/EC of the European Parliament and of the Council, Commission Directives 2002/67/EC and 2008/5/EC and Commission Regulation (EC) No 608/2004. OJ L 304 22.11.2011, pp 18–63

European Parliament and Council (2015) Regulation (EU) 2015/2283 of 25 November 2015 on novel foods, amending Regulation (EU) No 1169/2011 of the European Parliament and of the Council and repealing Regulation (EC) No 258/97 of the European Parliament and of the Council and Commission Regulation (EC) No 1852/2001). OJ L 327, 11.12.2015, pp 1–22

European Parliament and Council (2018) Regulation (EU) 2018/848 of 30 May 2018 on organic production and labelling of organic products and repealing Council Regulation (EC) No 834/2007. OJ L 150, 14.6.2018, pp 1–92

France

Agence nationale de sécurité sanitaire de l'alimentation, de l'environnement et du travail (ANSES) (2015) Opinion on the use of insects as food and feed and the review of scientific knowledge on the health risks related to the consumption of insects, ANSES Opinion Request No. 2014-SA-0153, 12 February 2015. Available at https://www.anses.fr/en/system/files/BIORISK2014sa0153EN.pdf. Last accessed 26 July 2020

Italy

Ministero della salute (2018) Informativa in merito all'uso di insetti in campo alimentare con specifico riferimento all'applicabilità del Regolamento (UE) 2015/2283 sui "novel food", 8 gennaio 2018. Available at http://www.trovanorme.salute.gov.it/norme/renderNormsanPdf?anno=2018&codLeg=62647&parte=1%20&serie=null. Last accessed 26 July 2020

Ministero delle Politiche Agricole, Alimentari e Forestali (MIPAAF) (2020) Ventesima revisione dell'elenco dei prodotti agroalimentari tradizionali. 20 February 2020. https://www.politicheagricole.it/flex/cm/pages/ServeBLOB.php/L/IT/IDPagina/15132. Last accessed 26 July 2020

Netherlands

Netherlands Food and Consumer Product Safety Authority (2014) Advisory report on the risks associated with the consumption of mass-reared insects, 15 October 2014. Available at https://zenodo.org/record/439001#.Xx1YF55KjnY. Last accessed 26 July 2020

Portugal

Costa MJ, de Moura Murta D, Oliveira Novais Leite de Magalhães T. Code of Good Practice for insect production, processing and use in animal feeding, Direção-Geral de alimentação e Veterinária (DGAV), 1 August 2018

Portugal Insects and Direção Geral de Alimentação e Veterinária (DGAV) (2019) Questions and Answers regarding the use of insects for human consumption. Available at https://en.portugalinsect.pt/faq. Accessed 26 July 2020

Switzerland

Conseil fédéral suisse (2016) Ordonnance sur les denrées alimentaires et les objets usuelsdu 16 décembre 2016, RS 817.02

Département fédéral de l'intérieur (2016a) Ordonnance sur les nouvelles sortes de denrées alimentaires du 16 décembre 2016, RS 817.022.2

Département fédéral de l'intérieur (2016b) Ordonnance concernant l'information sur les denrées alimentairesdu 16 décembre 2016, RS 817.022.16

Office fédéral de la sécurité alimentaire et des affaires vétérinaires (2017) Lettre d'information 2017/1: Production et transformation d'insectes à des fins alimentaires, 6 April 2017. Available at https://www.blv.admin.ch/blv/fr/home/lebensmittel-und-ernaehrung/lebensmittelsicherheit/einzelne-lebensmittel/insekten.html

Office fédéral de la sécurité alimentaire et des affaires vétérinaires (2020) Applications for placing on the market novel and novel traditional foodstuffs, 17 June 2020. Available at https://www.blv.admin.ch/blv/en/home/lebensmittel-und-ernaehrung/rechts-und-vollzugsgrundlagen/bewilligung-und-meldung/bewilligung.html. Last accessed 20 June 2020

United States of America

US Government (1938) Food, Drug and Cosmetics Act. Consolidated version available https://uscode.house.gov/browse/prelim@title21&edition=prelim. Last accessed 27 June 2020

US Government (1996) Code of Federal Regulations. Available at https://ecfr.io/. Last accessed 27 June 2020

Food and Drug Administration (1995) Defect levels handbook – the food defect action levels: levels of natural or unavoidable defects in foods that present no health hazards for humans. Available at https://www.fda.gov/food/ingredients-additives-gras-packaging-guidance-documents-regulatory-information/food-defect-levels-handbook#using. Last accessed 27 June 2020

Food and Drug Administration (2015) Presentation by Dr. George Ziobro (FDA Center for Food Safety and Applied Nutrition) at Institute of Food technologists (IFT) – Regulatory Issues, Concerns, and Status of Insect Based Foods and Ingredients. Chicago, IL

World Organisation for Animal Health

OIE Rift Valley Fever (2020). Available at https://www.oie.int/en/animal-health-in-the-world/animal-diseases/Rift-Valley-fever/. Last accessed 29 June 2020

Stakeholders' Materials

IPAA (2020a) Insects as food. Available at https://www.insectproteinassoc.com/insects-as-food

IPAA (2020b) Insects as feed. Available at https://www.insectproteinassoc.com/insects-as-feed

IPIFF (2019a) Animal welfare in insect production. Available at https://ipiff.org/wp-content/uploads/2019/02/Animal-Welfare-in-Insect-Production.pdf. Last accessed 10 July 2020

IPIFF (2019b) The European Insect Sector today: challenges, opportunities and regulatory landscape: IPIFF vision paper on the future of the insect sector towards 2030. Available at https://ipiff.org/wp-content/uploads/2019/12/2019IPIFF_VisionPaper_updated.pdf. Last accessed 10 July 2020

IPIFF (2019c) Regulation (EU) 2015/2283 on novel foods – Briefing paper on the provisions relevant to the commercialization of insect-based products intended for human consumption in the EU, V.2 Brussels, August 2019. Available at https://ipiff.org/wp-content/uploads/2019/08/ipiff_briefing_update_03.pdf. Last accessed 27 June 2020

IPIFF (2019d) IPIFF guide on good hygiene practice for EU producers of insects as food and feed. Available at https://ipiff.org/wp-content/uploads/2019/12/IPIFF-Guide-on-Good-Hygiene-Practices.pdf. Last accessed on 10 July 2020

IPIFF (2019e) Contribution Paper – EU organic certification of insect production activities, 29 March 2019. Available at https://ipiff.org/wp-content/uploads/2019/03/IPIFF_Contribution_Paper_on_EU_organic_certification_of_insect_production_activities_29-03-2019.pdf. Last accessed 10 July 2020

IPIFF (2020a) The insect sector milestones towards sustainable food supply chains, updated May 2020. Available at https://ipiff.org/wp-content/uploads/2020/05/IPIFF-RegulatoryBrochure-update07-2020-1.pdf. Last accessed 27 June 2020

IPIFF (2020b) Edible insects on the European market, June 2020. Available at https://ipiff.org/wp-content/uploads/2020/06/10-06-2020-IPIFF-edible-insects-market-factsheet.pdf. Last accessed 27 June 2020

IPIFF (2020c) IPIFF guidance – the provision of food information to consumers. Available at https://ipiff.org/wp-content/uploads/2019/09/FIC-doc.pdf. Last accessed 10 July 2020

PRoteINSECTS (2016) White paper: insect protein – feed for the future addressing the need for feeds of the future today. Available at http://www.proteinsect.eu/fileadmin/user_upload/press/proteinsect-whitepaper-2016.pdf. Last accessed 27 June 2020

Printed in the United States
by Baker & Taylor Publisher Services